Unlocking the Secrets of Commercial Properties: The FM Engineering Roadmap

Rise Above the Crowd

Angelo M Noto

Editing, design, typesetting and publishing
by UK Book Publishing.

www.ukbookpublishing.com

ISBN: 978-1-917329-17-0

Contents

This book gives you a comprehensive view of facilities management and its daily challenges. It also shows you how to prepare yourself and the team around you for the rapid change that the future brings. Along with sound technical advice, Angelo offers guidance on the equally, if not more, important tools for staying ahead of future environmental challenges. I'm happy to recommend this book for anyone who wants to help build and maintain a better tomorrow.

John Field - Author of "From engineer to Manager" and
Director at RapportTech Learning Solutions-

Introduction

I vividly recall one of the last days at school before summer break. Our teacher asked about our future plans. While many of my peers eagerly chimed in, I remained silent, listening intently. Deep down, I already knew my path. Not long after high school, I enlisted in the army. Following a few months of rigorous training, I embarked on my first tour in Kosovo. I won't delve into the specifics, but suffice it to say, my role didn't involve engineering only. It was an eye-opening experience for someone my age, teaching me the profound importance of values like discipline, honesty, empathy, sacrifice, and abnegation. Though the Kosovo tour was transformative, it paled in comparison to the challenge that awaited me in Afghanistan. By then, I was part of the engineering team in a specialized army division: The Parachuters. Again, I'll spare you the details, but it was undeniably hard. Upon returning from Afghanistan, I opted to transition from military life to civilian life, leveraging my skills in new ways. Initially, I felt like a fish out of water. However, I soon realized that over the years, I had unconsciously improved vital qualities: leadership, resilience, emotional intelligence, communication, humility, empowerment, and a passion for engineering. These qualities equipped me for success in my new chapter. More than a decade later, my pursuit for self-improvement persists. Among all the changes and challenges, one constant remained – my wife. Her influence has been immeasurable. She played a fundamental role in my career, continuously encouraging me to push forward. The saying "Behind every great man, there's a great woman" finds its truth in our life's narrative with the only difference that I don't see myself as a great man but she is definitely a great woman! She taught me how to be ambitious without descending into arrogance and instilled in me a belief in my own potential. In every failure, she offered support; in every win, she was there to celebrate. It is, in no small measure, thanks to her that I have matured into not only a better professional but, more importantly, a better person. Truly, in her and my daughter, I found my fortune; I was, and am, a lucky man.

I've put together this guide because I really want to share some lessons, insights, and a bit of advice that I think could really help you out if you're aiming to excel in this field and grab the title of "The Key Holder". Now, I know some of this stuff might seem either super interesting or a bit on the predictable side to you. But stick with me! What I'm really aiming for is to fill in those gaps that the formal education system often leaves wide open. This guide is your friend in getting to grips with the essentials every engineer should know. We're talking the daily tools and tricks of the trade, getting a 360 view on property management – covering finance and energy to taking the lead. It's all about keeping on learning and growing, because, let's face it, staying sharp and expanding our know-how is what keeps us ahead in any field. And when it comes to working well with others, it's all about clear communication and a good dose of patience. No matter if you're just starting out or you've been in the game for ages and are looking for a new spark, this book will guide you in identifying where to concentrate your efforts to stand out from the crowd. This isn't just about professional growth; it's about shaping you into a well-rounded engineer, a solid individual, and an awesome team player ready to take the lead on projects and people.

So, are you ready? Let's get started on this journey together.

Section
1

Who is "The key holder" and how to become one?

In the vast realm of the technical industry, amongst the intricate threads of innovation and design, there lies an often-understated element of distinction. This strand separates those who merely work from those who carry a profound responsibility, a sense of mission. I refer to these select few as "The Key Holders". But who is this figure, this keeper of keys, and how can you shape yourself into this model? Allow me to take you on a journey. Imagine walking into a dimension where the air buzzes with ideas and solutions. In this space, a Key Holder doesn't simply exist as someone with a specific skill set. No, they are the embodiment of a mindset. They don't walk the path of the ordinary; they craft their own, driven by the question, "how can I do it better?" rather than settling for a mere "that's okay". You might think that this passion consumes them, that they spend all their time on their craft. But here's where the Key Holder surprises you. They balance. They carve out moments for leisure, for reflection, and for their loved ones. After all, true inspiration often shoots in these quiet moments away from the routine. Now, leadership is an art, a combination of influence and inspiration and our Key Holders master it elegantly. They lead, not from a pedestal, but from the ground, walking hand in hand with their team. They wear their scars, their mistakes, not as badges of shame but as lessons learned. Their humility shines when they step back to admit a misstep or extend an apology. Yet, don't mistake their humility for compliance. They aren't mere spectators in the world of decision-making. They guide, they steer, even when it means challenging the currents. Their voice is not one of blind agreement; it echoes with knowledge and a desire to pave the right path. So, how do you become this model of excellence? The journey inward is as crucial as the journey outward. The world of engineering evolves with each tick of the clock, and the Key Holder moves with it, remaining greedy for knowledge. Every experience, every situation, no matter how small, is a page in their journal, a reflection of lessons learned and insights gained. Progress is a mosaic of tiny, incremental changes. It's about looking in the mirror every day and seeking growth, even if it's the smallest change. It's recognizing weaknesses and viewing them as fertile grounds ready for cultivation. And it's about stepping out, taking the reins, and challenging oneself, always with the burning question: "How can I do it better?" Yet, mechanical, electrical, carpentry, emergency situations, the Key Holder never forgets the human touch. Relationships are their bridge to greater impact. Every conversation, every interaction, is a step towards building trust, understanding, and mutual growth. In essence, becoming a Key Holder isn't a simple title to be achieved. It's a metamorphosis, a relentless pursuit of self-improvement and excellence. All you must do is embark on the journey. The door to greatness awaits, and remember, you have the power to hold the key and open it.

SUCCESS! How to get there?

In any organization or business, good engineers must have relevant skills; however, this is not enough – being a genuine person is a priority as it encourages creating a successful team. Also, it is essential to have personal values and personality – your values are literally the foundation for success. Understanding and possessing a willingness to work hard and work smart means learning the most efficient way to complete tasks and finding ways to save time, while completing daily assignments. It's also important to care about your profession and complete all projects while maintaining a positive attitude. Doing more than is expected is a good way to show that you have good management skills. Being at work on time doesn't need any skill; however, many fail on the very basic rules and this is a simple and essential task which helps to create your reputation inside a team. You will be there when you are supposed to be, keeping the team well informed of changes in your schedule or if you are going to be late for any reason. This also means keeping your team informed on where you are on the projects you have been assigned to. You can be dependable and responsible, also by keeping the team informed of any matters which may concern them, showing that you value your job and that you are capable of keeping up with the projects assigned. Get the work done and motivate others to do the same without focusing on the obstacles that inevitably come up in any job, is the right mindset. With a positive attitude it is possible to take the initiative and have the motivation to get the job completed in a reasonable period. You have the power to influence others with your behaviour. Working in a passionate team with positive attitude creates an environment of goodwill which is highly appreciated by anyone in your organization. Being open to changes and improvements provides an opportunity to complete work assignments more efficiently, while offering additional benefits to the team and guests. While frequently there are complaints in regards to changes in the workplace, which can be taken as they don't make sense or make the work harder, often these complaints are due to a lack of flexibility. Adaptability also means adapting to the personality and work habits of co-workers and leaders. Adapting your personal behaviour to accommodate others is part of what it takes to work effectively as a team; changes can be an opportunity to complete work assignments in a more efficient manner, adapting to change can be seen as a positive experience. Often, new strategies, ideas, priorities, and work habits can make the workplace more dynamic.

Above all, good relationships are built on trust. *"I want to know that I can trust what you say and what you do"* – that's what people think. Successful businesses work to gain the trust of customers; the same principle applies in your team and it is your choice to use your own individual sense of moral and ethical behaviour. You don't need supervision or direction to get the work done in a timely and professional way – once you understand your accountability on the job, you will do it without any prodding from others. Working in a supportive work environment and taking the initiative to be self-directed will provide you with a better sense of accomplishment and increased self-esteem.

Knowledge is power. It is very true! Keeping yourself updated with new developments and knowledge in the field, learning new skills, techniques, methods, and/or theories through professional progress helps to give you the confidence to think out of the box and it makes the job more interesting and exciting. Keeping up with current changes is vital for success and it increases opportunities. Self-confidence is the key ingredient between someone who is successful and someone who is not. A self-confident person is someone who inspires others. A self-confident person is not afraid to ask questions on topics where they feel they need more knowledge. You don't want impress others by imposing your knowledge, but confidence means instead feeling comfortable as it is not needed to know everything or be the most interesting person in the room. A self-confident person does what feels right and is willing to take risks, admitting mistakes. It's important to be aware of your strengths, but it is even more important to understand weaknesses, which allows you to have a clear idea on what to improve. Having faith in yourself and your abilities gives you a positive attitude and outlook on life. Another factor which helps you to thrive is professional behaviour, which includes learning every aspect of a profession and doing it to the best of your ability. Professionals look, speak, and dress accordingly, to maintain an image of someone who takes pride in their behaviour and appearance. Professionals complete high-quality work and are detail-oriented. Being professional includes all the above in addition to providing a positive role model for others. Professionals are enthusiastic about their work while being optimistic about their organization and future. To become a professional, you must feel like a professional, and following these tips is a great start to achieve your goal and success.

The importance of prioritising tasks

As engineers, we often find ourselves juggling multiple projects, tasks, and deadlines. It can be tempting to try to do everything at once, believing that multitasking is the most efficient way to get things done. However, the reality is that trying to do everything chaotically can lead to mistakes, missed deadlines, and a lack of focus on what's truly important. The key to success is learning how to prioritize tasks effectively. This means taking the time to evaluate each task based on its importance, urgency, and level of effort required. By doing so, you can make informed decisions about which tasks to focus on first, which to delegate, and which to postpone or eliminate. Prioritising tasks is crucial for several reasons. First, it helps you avoid the stress and overwhelm that can come from trying to do too much at once. When you have a clear sense of what needs to be done and when, you can approach each task with focus and intentionality, knowing that you're working on the most important things first. Second, prioritising tasks helps ensure that you're meeting your deadlines and delivering high-quality work. You are focusing on the most critical tasks first, which makes it less likely important deadlines slip, and you'll have more time and energy to devote to ensuring that your work is of the highest quality. On the other hand, trying to do everything chaotically can lead to myriad risks and negative outcomes. For example:

- You may miss deadlines or deliver poor work because you haven't given yourself enough time to focus on each task properly.
- You may feel constantly overwhelmed and stressed, leading to burnout and reduced productivity over time.
- You may struggle to prioritize effectively, leading to a situation where you're always working on the tasks that are most urgent, rather than those that are truly important.
- You may miss out on opportunities to delegate tasks to others or collaborate with colleagues, leading to a lack of teamwork and missed opportunities for growth and learning.

In conclusion, prioritising tasks is a critical skill for any engineer who wants to be successful and productive. I have always taken the time to evaluate each task, prioritise based on importance and urgency, and focus on one thing at a time. In this way I try to have a clear idea of the situation and I can ensure that I'm delivering the most high-quality work I possibly can, on time, and having a clear action plan, avoiding the risks of trying to do everything confusedly.

The power of writing

As a professional, you are responsible for overseeing projects, delegating tasks, and ensuring that everything runs smoothly. With so much going on, it can be easy to forget details and lose track of important information. That's why it's crucial to take notes during meetings and conversations. Writing things down gives force to your words and helps you remember important details. It's also an efficient way to keep track of tasks, deadlines, and progress. Here are some reasons why you should consider note-taking a priority. Writing things down engages your brain in a different way than just listening or speaking. It requires more focus and attention, which can improve memory and recall. When you write down important information, you are more likely to remember it later. This is especially important for managers who need to keep track of multiple projects and tasks. Writing can also help you process information and gain a deeper understanding of complex topics. When you take notes, you have to think about the information and rephrase it in a way that makes sense to you. This can help you clarify your thoughts and identify areas where you need more information or clarification; for example, taking notes during meetings and conversations can also improve communication and collaboration. When everyone has a clear record of what was discussed and agreed upon, it's easier to follow up on tasks and hold people accountable. It can also prevent misunderstandings and conflicts by ensuring that everyone is on the same page. Finally, taking notes can help you make better decisions recording information and ideas – you can review them later and use them to inform your choices. You can also use your notes to identify patterns and trends that can help you make predictions; in fact, it is a simple but powerful tool that can help anyone to be more efficient, productive, and effective. Making it a habit can improve your communication, organisation and decision-making skills. Don't underestimate the power of writing – it can be a game-changer for your management style.

Planning ahead

Planning ahead is a crucial aspect of effective management. A professional who plans days ahead can increase efficiency, reduce stress, and achieve better results. In this section, we will discuss the benefits of planning ahead and provide tips on how to do so effectively. When you plan your days ahead, you can prioritize tasks and allocate time accordingly. By doing so, you can ensure that important tasks are completed first, and that there is enough time to complete all tasks accordingly. This can increase efficiency, reduce wasted time, and help to achieve more in less time. Planning ahead can also help to reduce pressure. When, for instance, I have a clear plan for the day, I can feel more in control of the situation and less overwhelmed; this can lead to a greater sense of calm and focus, which can ultimately lead to better decision-making and improved performance. Additionally, potential obstructions can be identified and develop strategies for overcoming them.

Tips for effective planning

A planner or scheduling tool can help to stay organized and on track. There are many different options available, from physical planners to digital apps. I personally found both options very helpful depending on the location or type of project; however, I always prefer to have everything in once place. Maintaining realistic goals for you and your team can help to avoid overloading schedules and reduce anxiety.

It's important to build in flexibility when planning ahead because unexpected events can arise, so it's vital to have a contingency plan in place and be prepared to adjust the schedule if needed.

Achieving your goals

Achieving your goal and dreams is a common aspiration for many people, but it requires discipline and consistency to make those aspirations a reality. Below are some bullets points which helped me create a successful career path.

- The first step in achieving your goals is to set specific and measurable goals. Specific goals are clear and concise, and measurable goals allow you to track your progress and stay motivated. When setting your goals, make sure they are realistic and achievable, but also challenging enough to motivate you.
- Once you have written your goals, create a plan that outlines the steps you need to take to achieve them. Your plan should include specific actions you need to take, as well as deadlines for each action. This will help you stay on track and ensure that you are making progress towards your goals.
- One of the biggest challenges in achieving your goals is staying focused. It's easy to get distracted by other things in your life, but it's important to stay focused on your goals. One way to do this is to prioritise your goals and make them a priority in your daily life. This means dedicating time and energy to working towards it every day, even when it's challenging.
- You need to stay motivated. Motivation can come from many sources, such as positive feedback from others, visualising your success, or rewarding yourself for making progress. Find what works for you and use it to stay motivated throughout your journey.

- Discipline and consistency are essential as it means sticking to your plan, even when it's challenging, and taking consistent action towards your goals every day. It also means holding yourself accountable and staying committed to your goals.
- Delays are a natural part of any journey, and it's important to learn from them. When you encounter a hold-up, take the time to reflect on what went wrong and how you can adjust your plan to avoid similar situations in the future. Remember, impediments are not failures, but opportunities to learn and grow.

Staying up to date

Always being up to date is crucial for success in any industry, but it is especially important in the engineering sector. The engineering sector is constantly evolving, with new technologies and innovations emerging all the time. It's important to stay up to date with these emerging technologies, as they can have a significant impact on the industry and help you to identify new opportunities and stay competitive in the market. You can spot new areas for professional development and seek out opportunities to improve your skills. This can lead to increased job satisfaction, higher pay, and better career opportunities. In addition, regulations and standards are constantly evolving, and failing to stay up to date can result in costly mistakes, legal problems, or even injury or loss of life. Keeping up with the latest stuff is super important for those who want to stay ahead of the game, you can find new chances to shine and get ahead of others. This means you could do even better in your job and have more opportunities come your way. One great way to stay in the loop is reading specialised magazines and websites. They're full of the latest news and trends. When you read these, you're always in the know about what's happening. Also, joining groups can be a big help as you learn lots of useful stuff and meet people who can help you out. I found the organizations listed below absolutely fantastic and most useful:

- CIBSE (Chartered Institution of Building Services Engineers) https://www.cibse.org/membership-registration
- BESA (Building Engineering Services Association) https://www.thebesa.com/besa-member-area/technical
- IWFM (Institute of Workplace and Facilities Management) https://www.iwfm.org.uk/membership.html
- BSRIA (Building Services Research and Information Association) https://www.bsria.com/uk/membership/membership-benefits/
- CPD UK (Continuing Professional Development) https://cpduk.co.uk/courses?category=engineering
- EMA (Energy Management Association) https://www.theema.org.uk/ema-membership/

Attending conferences, workshops and webinars are also great opportunities to learn about new technologies, networking, and gain new insights into emerging trends. Many companies also offer training and development programmes.

A "Yes man" is not a successful man

In our world, success is often defined by the ability to solve problems and make decisions that benefit the company. I believe that agreeing with everything the superiors say is therefore not the key to success. I can confirm that being a "yes man" is not a successful approach in any profession. When someone is a "yes man", they are often viewed as lacking in independent thinking. Instead of critically analysing situations and providing their own insights, they simply agree with whatever they are told and this can lead to missed opportunities and poor outcomes. If we blindly follow instructions without questioning their validity, it's easy to end up making errors that could have been avoided, so it is crucial to have people on the team who can think critically and independently. If not, it can halt career progress and lead to stagnation. This might also lead to a loss of trust and respect from colleagues. Being viewed as someone who simply agrees with everything could suggest you're not an essential team member but instead someone who could be easily replaced. Below are some tips for avoiding the "yes man" trap

Ask questions: Don't be afraid to ask questions and seek clarification when necessary, so that you can gain a deeper understanding of the problem at hand and identify potential solutions that may not have been considered.

Provide your own insights: Instead of simply agreeing with everything you are told, provide your own insights and ideas and you can demonstrate your independent thinking skills and show that you are a valuable member of the team.

Challenge assumptions: Don't be afraid to challenge assumptions and conventional wisdom as you can help identify new solutions and approaches that may have been overlooked.

Be confident in your abilities: Have confidence in your abilities and trust in your own judgment. This can help you avoid the temptation to simply agree with everything you are told and instead make decisions that are in the best interests of the company and its customers.

Trust vs Performance

One of the most important decisions that you can make during your career is choosing between trust and performance when evaluating a talent. On the one hand, someone who is highly skilled and productive can be a valuable asset to the team. On the other hand, someone else who lacks integrity and cannot be trusted may cause problems down the line. It's complex but we can try to explore how to navigate this decision by using a trust-performance matrix.

The matrix is divided into four quadrants:

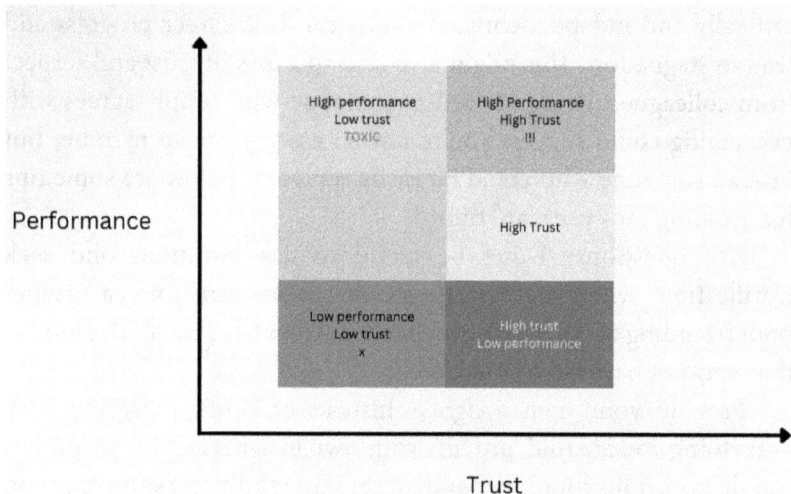

1. High trust, high performance: all of us want them, the ideal candidates for a leadership position. They are highly skilled and productive, and they can be trusted to act in the best interests of the company.
2. High trust, low performance: They may not be the most productive or skilled, but they are highly trustworthy. They may benefit from additional training and support to improve their performance.
3. Low trust, high performance: Here is where it becomes interesting as they may be highly skilled and productive, but they lack integrity and cannot be trusted.
4. Low trust, low performance: These employees are not productive or trustworthy. They may need to be let go or reassigned to a different role.

We should always consider both trustworthiness and performance. We should fight for those who fall into the high trust and high-performance quadrant – you don't want to lose them, they are rare, and leaders should proactively consider promoting them to leadership positions or giving them additional responsibilities, as they have demonstrated their value to the company and can be trusted to act in the best interests of the team. For the high trust, low performance quadrant, consider providing additional training and support to help improve skills and productivity. They have demonstrated their trustworthiness and may be valuable assets to the team with additional support.

The low trust, high performance quadrant, in this case closely monitor the behaviour and address any issues that arise. Valuable in the short term, but a lack of integrity can cause issue in the future. Some time and effort should be invested to improve behaviour and build trust. Finally, the low trust, low performance quadrant, often the best decision is to consider letting them go to avoid a negative influence on the team.

It's clear that both high performance and trust are desirable qualities; however, I personally always choose trust over performance. As Herb Kelleher once said: *"We will hire someone with less experience, less education, and less expertise, than someone who has more of those things and has a rotten attitude. Because we can train people. We can teach people how to lead. We can teach people how to provide customer service. But we can't change their DNA"* and I'm absolutely in agreement with that. I believe that lack of honesty can cause complications such as cutting corners or engaging in unethical behaviour to achieve the goals, which can harm the team's long-term success. In contrast, a talent can be highly trustworthy but their lack of skills can be trained and developed to become a more productive team member because a team that trusts each other can work more efficiently together. Highly skilled but a lack of trustworthiness can disrupt team cohesion and cause tension and mistrust among team members.

Another important aspect is the "Reputation", the leader's reputation, the department's reputation and the company's reputation. Here it is another of my favourite quotes from Halford E. Luccock which says *"No one can whistle a symphony, it takes an orchestra to play it"*; this is the school of thought which I like.

The Risks of Job-Hopping

In your career, you'll find lots of attractive job options that might tempt you to change jobs often. You're not alone in this. The modern employment landscape, brimming with career prospects, often encourages this tendency. However, let's pause for a moment and consider this – is job-hopping always the best choice for your career? In this chapter, we'll explore the potential risks of job-hopping, the art of patience, and the importance of seeing the bigger picture beyond immediate financial gains. You might have a stellar set of skills, an exceptional work ethic, and impressive qualifications, but frequent job changes could paint a different picture on your CV. Potential employers often perceive a series of short-term roles as a sign of unpredictability or lack of commitment. Remember, companies invest significant resources into hiring and training new employees. What they're looking for is someone who would stick around and yield returns on this investment. Your job-hopping history might make them hesitate, wondering if you'd leave before they've really gained anything from your tenure. At the start of your career, it's tempting to chase after the highest salary. That's understandable; we all have bills to pay and life goals to save for. But if there's one piece of advice I want you to take to heart, it's this: focus more on the opportunities and less on the immediate remuneration. The beginning of your career is the time for learning, for growth, for building that robust professional foundation. Salary, while certainly important, should not be your sole deciding factor. The skills, experience, and exposure you gain during this period can lead to greater financial rewards in the long run.

You've accepted a job, and it's your first day. This is your chance to make a difference. But hold on a second! Rather than charging in and making sweeping changes, your priority should be to understand the building and its daily operations. Watch, listen, learn. Find the weak points that need improvement. Once you've gathered enough information, approach your team with your insights. Implement changes gradually and always ensure these are team decisions. By doing so, not only do you make sustainable improvements, but you also foster a team spirit, which will greatly enhance your work environment. The point is, you need time to make an impact, to demonstrate your capabilities and to thrive. A few months is not enough. So, as you journey through your career, remember that patience is not just a virtue, it's a strength. Focus on your long-term growth, engage fully in each role, and approach change methodically. And most importantly, don't forget to enjoy the journey while you're at it!

Let's focus for a moment on a crucial aspect which is "welcoming a new member" to your team and the importance of their first week. Just as we've discussed the significance of your approach when you start a new job, this is the flip side of the same coin. Imagine this: You've been given the responsibility to welcome a new team member. It's not only about a new hire; it's about setting the stage for a productive working relationship, and trust me, the first week plays a pivotal role in this. The newbie needs to observe, learn, and gradually adapt to their new environment, and it's equally important for you and your team to create a conducive atmosphere that aids this transition. A warm welcome can go a long way. From simple gestures like a welcome note, a team lunch, or a one-to-one introductory meeting, make sure you make your new team member feel comfortable and included right from the start. But remember, an effective introduction goes beyond mere pleasantries. This is where planning comes into the picture. Organizing a detailed and comprehensive training week is vital – this is your opportunity to showcase the professionalism and high standards you've set in your department. The first week's training should be designed with meticulous attention to detail. Ensure it's organized, clear, and comprehensive, covering all aspects of the new role and responsibilities. Include introductions to the tools and software they'll be using, the processes and protocols of your department, and the broader company culture and values. This well-planned training will not only make the newcomer more comfortable but also boost their confidence. They'll appreciate the structured introduction to their role, which can set the tone for their entire tenure in your team. Seeing the level of organization and professionalism you've implemented in your team's training process will give him a clear understanding of the standards and expectations that he needs to meet. So, while you focus on your personal career development, remember the role you play in your team's growth as well. As a leader, you're not just shaping your own future, but you're also influencing the careers of those who come after you. As you welcome new team members and help them navigate their first week, you're creating a strong, cohesive unit that's capable of accomplishing great things. After all, engineering isn't just about individual brilliance; it's about creating a team.

The Hidden Costs of Recruitment

Research from the CIPD reveals that the average cost of hiring the wrong person for a position can be around £12,000, taking into account just the cost of recruitment. This figure does not include expenses such as initial training, loss in productivity, or recruiter fees.

(CIPD stands for the Chartered Institute of Personnel and Development. It is a professional body for human resources and people development, with more than 150,000 members globally. The CIPD aims to promote the HR profession and advance the field of people management by providing research, training, and development opportunities to its members. The organization also advocates for policies that support fair and effective workplaces, and produces a range of publications, events, and resources for HR professionals and businesses.)

The REC conducted a similar study including the total cost of hiring the wrong person. Costs can exceed £40,000 per annum.

(REC stands for Recruitment and Employment Confederation. It is the professional body for the UK's recruitment industry, with more than 3,500 member companies ranging from small independent recruiters to large recruitment agencies. The REC provides training, support, and guidance to its members, and works to promote best practices in recruitment and employment across the industry. The organization also lobbies the government on behalf of the recruitment sector and conducts research and analysis on employment trends and practices.)

The above data is a clear and precious sign for leaders to take into consideration not only the upfront costs of recruitment, such as advertising, marketing, and agency fees, but also the hidden costs such as the man-hours spent on the recruitment process and, most crucially, the cost of hiring the wrong person as mentioned earlier. This is especially true when recruiting for a senior-level position. In today's skill-short market, it's necessary to go beyond just offering a job; in fact, marketing your company effectively to attract top talent is essential and you can do this by providing a clear career path that allows for building on skills. This requires careful consideration of factors such as salary (a good salary analysis of your area will help you to demonstrate the best salary offer for that specific position to the business executive team), benefits package, job description, development opportunities, and company ethos. Effective marketing is also key, so with businesses needing to reach as wide an audience as possible the traditional methods such as placing an advert in the local websites are no longer enough – job vacancies need to be marketed through a variety of channels to ensure they reach the right audience even in different countries. By taking a comprehensive approach to recruitment, you can stand out from the crowd and attract the best candidates. When discussing job hunting with senior-level colleagues, I've found that there are common themes in what they seek in a new employer. Regardless of career level or position, high-level roles often express dissatisfaction with their current job due to a lack of career progression, learning opportunities, salary growth, flexibility, challenge, or a general feeling of stagnation. It makes me think that if I want to be successful in attracting top talent, I need to be able to offer solutions to these issues. The first thing you can do is to write down a "providing opportunities list" for career development and learning, ensuring competitive salaries and benefits, offering flexible work arrangements, and creating a challenging and stimulating work environment. In this way you and your company can position yourselves as attractive options for the right candidates seeking a change. Here below are some useful topics for you to use:

- Create a positive and inclusive work culture.
- Provide challenging and innovative projects.
- Emphasize the company's impact and contribution to society.
- Ensure work-life balance and flexibility.
- Provide opportunities for growth and advancement within the company.
- Communicate clearly and effectively about the company's values and mission.

Try focusing on what the business can offer rather than just what the candidate can bring to the company; in this way you can attract the best in the market. It's a two-way road, and by creating a mutually beneficial relationship, both the business and the new talent can thrive.

Retention has become a very stressful matter, and I can ensure you, I know that feeling! I've learned that any leader who masters the skill of understanding what keeps top talent happy in their jobs will greatly succeed in retaining them. Studies, such as one conducted by Oxford University, show that happy employees are around 13% more productive, and are more than twice as likely to stay in a role where they are happy. This means not only retention, but also more productivity, leading to a win-win situation for both the talents and the employer. To keep top engineers happy, it is important to offer opportunities for personal and professional growth, such as in-house and external training, and provide a clear career path rather than just a job.

Additionally:

1. paying them slightly more than the average industry salary
2. making them feel like they are making a difference in their job
3. offering regular reviews and support

These are the crucial factors for retaining top engineering talent in your Dream Team. It is also important to learn from mistakes, but unfortunately, only 50% of the employees who leave their positions are given structured and scripted exit interviews, meaning that many companies do not understand why they have lost their employees. This can lead to the same mistakes being made with the next hire, perpetuating the cycle of turnover. Companies that take the time to understand what went wrong with previous employees and use that information to improve retention will have a better chance of keeping their high-quality staff. It is important to conduct full exit interviews with departing engineers, asking questions such as "What could we have done differently?" It is also important to understand what package the employee is moving to, whether they had other companies interested in them, and whether their skillset is in demand. Ultimately, I want to highlight that it is vital to understand that recruiting the best personnel requires time and effort. If you're not willing to put in the necessary effort, you will miss an opportunity and those skills will go elsewhere; and without a top team, you and your department won't be able to offer the best service. The results? Managing the team and the daily workload would become more difficult, leading to increased time wastage and general frustration.

Quality Means Doing It Right When No One Is Looking

The quote, "Quality means doing it right when no one is looking," which is attributed to Henry Ford, has always been, for me, a powerful reminder of the importance of integrity and commitment to excellence. We all know that in any industry the client's satisfaction and seamless experiences are paramount, therefore maintaining high standards can be the key to success and standing out among the competition. In this chapter, I will share my personal experiences and insights on why I understood this philosophy to be essential for thriving. Throughout my career, I have always taken great satisfaction in knowing that I have done a good job, even when working alone or without direct supervision. This sense of pride in my work has motivated me to consistently deliver high-quality results, which in turn has prevented me from having to revisit the same issues multiple times. This attention to detail and dedication to quality is vital for ensuring that guests have a comfortable and enjoyable stay with us. If we striving for excellence in every task, no matter how small, we can create a great experience for our guests and contribute to the overall success of the business which will help to avoid or mitigate those very annoying comments and feedback in regards to the state of the facilities and their functionality. Our work style and approach to quality often serve as our signatures, reflecting who we are and what we value. Consistently delivering high-quality results can create a positive reputation not only for us but also for those who may eventually take our positions. By setting a strong example and maintaining a commitment to quality, we can inspire future talents to uphold the same high standards, fostering a legacy of excellence within the organization. This approach not only benefits the individuals involved but also contributes to the overall success and reputation of your career. One key aspect of ensuring quality is using the most effective techniques and tools for the job. I'm obsessed with investing my time and effort into learning best practices and mastering the appropriate tools, because in doing so I can improve efficiency and produce better results. For example, during my time as an engineer, I've discovered that using the right tools can significantly reduce the time of work and the likelihood of equipment breakdowns and extend the lifespan of essential systems. This attention to detail and commitment to using the proper tools have helped me consistently deliver high-quality work and maintain good guest satisfaction. In order to excel, it's crucial to continually educate ourselves and seek out the best solutions to the challenges we face. Dedication to learning can lead to developing the skills and expertise necessary to stand out in our field. In my experience, this commitment to learning and personal growth has not only led to improved job performance but has also instilled a sense of pride and passion for the work I do. A strong commitment to quality can also have a positive impact on our relationships with supervisors and other leaders within the organization. When we consistently deliver high-quality work and demonstrate a dedication to excellence, we can build trust and confidence in our abilities. Over time, this trust and confidence can lead to increased opportunities for growth, development, and advancement within the organization. Remember! Always try for quality, even when no one is looking; you can then create a strong foundation for success. The commitment to do good can also have a positive influence on our team members – we can inspire those around us to adopt the same mindset and approach to their tasks. In my experience, leading by example has resulted in a more cohesive and focused team. When team members understand the importance of doing things right, even when no one is looking, they are more likely to take ownership of their responsibilities and work together to achieve common goals. This shared commitment to quality can lead to a more positive working environment, increased team morale, and a stronger sense of camaraderie. There is always room for improvement, and it is essential to remain adaptable and open to change. Embracing a mindset of continuous improvement is the right direction for consistently enhancing our skills, knowledge, and overall performance. One effective approach to continuous improvement is to regularly assess and evaluate our work, seeking feedback from colleagues, supervisors and friends. This feedback can provide valuable insights into areas where we may need to refine our techniques or improve our understanding of specific systems and processes. Seeking out opportunities for improvement and using the feedback, especially the negative ones, will help maintain and ensure that we are always striving for excellence. The importance of quality cannot be overstated. If we consistently deliver high-quality work, we can gain numerous benefits:

1. Enhanced guest satisfaction
2. Improved operational efficiency
3. Increased opportunities for advancement
4. Positive team dynamics

I can't stop repeating it: "Quality means doing it right when no one is looking".

This philosophy will help you to rise above the crowd, contribute to the success of your organization, and create a lasting legacy of excellence within the industry. I hope that sharing my personal experiences and insights will inspire you to adopt a similar mindset and approach.

With dedication, continuous learning, self-motivation, you can achieve great things.

Do You Know That Sense of Satisfaction When You Fix Something?

Have you ever felt the unique joy that goes through you when you've managed to put together the scattered pieces of a puzzle? Or the euphoria when you've corrected a mistake in a complicated maths problem? I am fortunate enough to experience this profound sense of satisfaction. It's not just about fixing things; it's about restoring harmony, reinstating functionality, and often, breathing new life into machines that form the backbone of the building. Many see a faulty machine and immediately think of the inconvenience it causes. They might see a non-operational elevator, a faulty HVAC system, or even a machine halted in its operation. But when I look at these, I don't just see malfunctioning machines – I see challenges, puzzles waiting to be solved. Behind every blinking error light, there's a story, a trail of missteps, wear, or unexpected interruptions that led to this moment. When I approach a broken machine, it is like to a conversation. I begin by asking it, to tell me what's wrong. The initial symptoms offer clues, and each machine with its unique design and function, presents a new narrative of malfunction, wear, and tear. The journey from identifying the problem to rectifying it's not always straightforward and of course it would happen on Fridays or at the weekend… but this is another story. There are times when I find myself in a maze of complexities, where one problem leads to another or where the solution seems impossible, but, within lies the thrill. The more intricate the problem, the more rewarding the resolution. Why is this sense of satisfaction so intense, so deeply ingrained in our spirit? I believe it's because, at our core, humans are problem-solvers. From the days of our earliest ancestors, who innovated tools to make life more manageable, to today, where we troubleshoot complex technological marvels, we have thrived on our ability to identify problems and forge solutions. From a technical perspective, my work is not just practical, it's deeply human. Every time I manage to fix something, I'm not just restoring a machine; I'm ensuring that life goes on uninterrupted for countless individuals. Whether it's ensuring a hotel's critical equipment runs smoothly or fixing an electrical issue, my role impacts lives. That weight, that responsibility, amplifies the satisfaction of every successful repair and I wish I could bottle up this feeling and share it with the world. If more people understood the joy in rejuvenation, in restoration, perhaps we'd live in a world more inclined to fix what's broken rather than discard it. In a society that often prioritizes the new, the shiny, the innovative, I find profound contentment in preservation, in ensuring that what we have continues to serve us well. So, the next time something breaks down and you fix it, I invite you to take a moment and catch a fleeting sense of that profound feeling of joy, because in a world full of things that break, there's a silent heroism in being the one who fixes them. As with many vocations, the route of an engineer is one of growth, adaptation, and evolution. In the early stages, the thrill often stems from the tangible, such as, the turning of a spanner, the replacement of a faulty part, the physical act of repair. It's raw and immediate. But as one progresses, the nature of this satisfaction metamorphoses. It becomes less about the tangible fixes and more about understanding the intricacies of complex systems, the interplay of components, and the broader impacts of effective maintenance on organizations and society. In the early phase of my career, I often worked closely with equipment, the hands-on interaction was my primary mode of operation. The immediate feedback of a machine buzzing back to life under my touch was electrifying As I ascended the professional ladder, I began to distance myself from the ground level, not due to a weakening love for the craft, but as a natural progression towards the macro elements of engineering systems. Suddenly, it wasn't just about understanding how one machine functioned, but how an entire building operated. It became a game of optimization. How can we improve efficiency? How can the coordination of different systems result in overall better performance? How can the hard disciplines, like mechanical and electrical engineering, synergise with the softer elements, like Health and Safety, security, cleaning, guests' relations, to ensure that a building isn't just functional but also beneficial to its occupants? The satisfaction I derived began to stem from understanding these holistic systems. When I could propose a solution that not only rectified a singular problem but improved the overall efficiency, sustainability, or comfort of a building, the joy was immeasurable. It was similar to solving a multi-layered puzzle where each piece affected the whole picture. Don't get me wrong, the transaction period wasn't easy: at some point I missed handling the tools and being part of the team in the field, I was questioning myself if I was still good enough; but I understood that I have to make a choice, decide between enjoying the practical profession or embracing the fact that if I want to progress, I have to accept to change my way of working, responsibilities, how to maximise my time and how to strategically prioritise tasks. This new phase brought its challenges. Whereas before, the issues I encountered were often physical and immediately evident, now they were sometimes abstract, hidden and more sophisticated. It was about making things better, not just making them work. Moreover, as I progressed, the soft disciplines started playing a pivotal role. Understanding human behaviour, predicting how users interact with spaces, and even delving into areas like psychology became integral. After all, what's the use of a very efficient building if it doesn't make the people living in it comfortable and happy? In the initial stages, the wins were loud – machines roaring back to life, lights turning back on while in this advanced stage, the triumphs were silent. A building running at optimal efficiency, users seamlessly interacting with their environment, and systems working in perfect harmony – these became the new markers of success. Your journey, much like the equipment we work with, is one of continuous evolution. The satisfaction we derive from our work transforms as we advance, moving from the tangible to the conceptual, from the singular to the holistic, but throughout this journey, one thing remains constant: the indomitable spirit of problem-solving, the insatiable desire to improve, and the happiness from seeing systems, functioning at their very best. Understanding operation costs, maintenance and energy strategies are crucial to operate any property. POMEC is a set of calculation and schemes which helps to successfully manage properties' risks and mitigate costs, avoiding major complications.

The Operation Costs can be listed as below:

- Maintenance/Security/H&S
- Electricity, Gas, Water, Waste
- Insurance
- Personnel

If we don't maintain the facility, it will degrade, affecting health and safety of occupants and we will land into major repairing expenses at a later date. It is also essential to maintain its good appearance and functioning at maximum efficiency. It's clear that the speed of rate of decay and deterioration will depend on building design, quality of materials, workmanship, its main function, location and workforce. For these reasons, planned maintenance, preventive maintenance and corrective maintenance costs will be necessary for maintaining the property in good and habitable condition over its lifespan. We should take into consideration major repairs, emergencies and Mechanical, Electrical & Public Health (MEP) equipment replacements which should be executed over a specific period of time. These type of costs are generally called CAPEX or OPEX (Capital Expenditure and Operational Expenditure).

The most common maintenance costs are listed below:

- Minor repairs
- Painting
- Waterproofing on terrace and roofs
- Pest control
- Door-window repairs including fixtures, glass, locks
- Replastering or plaster repairs
- Plumbing repairs including change fixtures
- Training
- Paving repairs
- Furniture repairs
- Ventilation Systems including cleaning filters
- Statutory services
- Heating and Cooling systems
- Preventive maintenance
- Landscaping and Garden Maintenance

As we know implementing all kinds of active and proactive maintenance operations and energy control in a proper way can be challenging, the technology therefore can help to achieve the required targets thanks to unique and personalized software for professional preventive maintenance programmes measuring the machinery performance as well as the maintenance crew specifying, for example, the cost of maintenance against the initial installation cost and much more. Alternatively, and taking into account the current industry's trends, it is still common to use Excel spreadsheets.

Management

Businesses have a variety of expenses, from the rent they pay for their place of work or offices, to the cost of raw materials for their products; from the wages they pay to their workers to the overall costs of growing their business. To simplify all of these costs, businesses organise them under different categories. Two of the most common are Capital Expenditures (CAPEX) and Operating Expenses (OPEX).

CAPEX

Capital expenditures are purchases of significant goods or services that will be used to improve a company's performance in the future. Capital expenditures are typically for fixed assets like property, plant, and equipment (PP&E). For example, if a hospitality company will buy a new and more efficient set of boilers, the transaction would be a capital expenditure. One of the defining features of capital expenditures is longevity; meaning the purchases benefit the company in the long term.

The following are common examples of capital expenditures:

- Engineering plants
- Equipment and machinery
- Building improvements

Each industry might have different types of capital expenditures. The purchased item might be for the expansion of the business, updating older equipment, or expanding the lifespan of an existing fixed asset.

OPEX

Operating expenses are the costs a company incurs for running its day-to-day operations. These expenses should be ordinary costs for the industry in which the company operates.

The following are common examples of operating expenses:

- Wages and salaries
- Accounting and legal fees
- Overhead costs such as selling, general, and administrative expenses (SG&A)
- Business travel

Because of their different attributes, each is handled in a separate manner. OPEX are short-term expenses and are typically used up in the accounting period in which they were purchased. This means that they are paid weekly, monthly, or annually. CAPEX costs are paid upfront all at once. The returns on CAPEX take a longer time to realize; for example, machinery for a new project. The returns of OPEX are much shorter, such as the work that an employee does on a daily basis to earn their wages. If a company chooses to lease a piece of equipment instead of purchasing it, as a capital expenditure, the lease cost would be classified as an operating expense.

Understanding Utilities

Utilities are really important for all of us. They give us the things we really need every day like water, gas, electricity, taking away our rubbish, and keeping our phones and internet working. These are all part of what we call the public utility market. Some of these businesses use huge networks of wires and pipes to deliver electricity or gas straight to our home and buildings. Once these systems are set up, it doesn't really make sense to have other companies try to do the same thing in the same place. That's because it's cheaper and easier when one company does it for everyone. Some companies give us more than one thing, like both electricity and gas, while others focus on just one type of commodity, like water. Moreover, the modern public utility landscape is increasingly moving towards sustainable and renewable energy sources, with wind turbines and solar panels becoming common sights across the country. Below is a list of the key players on utilities distribution:

Generators: These are the companies that make the electricity or collect the water we use.

Network Operators: These folks manage the big networks of wires and pipes that bring the electricity or gas to our homes.

Traders and Marketers: They buy and sell commodities, sometimes making special deals for businesses like hotels to get what they need at a good price.

Service Providers and Retailers: These are the companies that sell the services directly to us, the customers, and you can choose which company you want to buy from.

Given their pivotal role in society, public utilities are bound by a social responsibility to:

- Ensure that their services are of the highest quality and responsive to the needs and desires of their users.
- Target health services effectively to enhance the health outcomes of local populations.
- Seek continuous improvements in service efficiency, ensuring that utilities operate sustainably and economically.

We typically receive utility bills from our suppliers on a monthly or quarterly basis. A proportion of your gas and electricity bill is used to fund the government's environmental initiatives.

Organizations have several options for how they purchase their utilities like gas and electricity. The type of contract they choose can significantly impact their utility costs and budgeting. Let's break down the differences between fixed, flex, and cash-out contracts:

Fixed Contracts lock-in the price you pay for your utilities for a specific period. This means the rate you pay for each unit of gas or electricity remains constant, regardless of market fluctuations.

Pros:

Budget certainty: Businesses can predict their utility costs, making budgeting easier.

Protection from price spikes: If utility prices go up, you still pay the agreed-upon rate.

Cons:

Missed savings: If prices fall, you're stuck paying the higher rate.

Commitment: You're locked in for the duration of the contract, which could be a disadvantage if your business needs change.

Flex contracts allow businesses to purchase utilities at market prices, which can vary. You can buy your utilities in advance (forward buying) or at current market prices, giving you the flexibility to take advantage of price changes.

Pros:

Potential for savings: If market prices drop, you can secure lower rates.

Flexibility: You can adjust your buying strategy based on market conditions and your business needs.

Cons:

Risk of price increases: If prices rise, so do your costs.

Complexity: Requires more active management and understanding of the market.

Cash-out contracts are a bit different and relate more to how imbalances in your utility consumption are managed. If you use more or less energy than you've contracted for, the "cash-out" rates are the prices you pay or receive for these imbalances.

Pros:

Flexibility in consumption: Useful for businesses with unpredictable energy needs.

Opportunity to benefit: If you can manage your consumption effectively, there could be financial benefits.

Cons:

Risk of high costs: If your consumption is much higher or lower than planned, the cash-out rates can be very expensive.

Market volatility: Prices are subject to market conditions, which can increase financial risk.

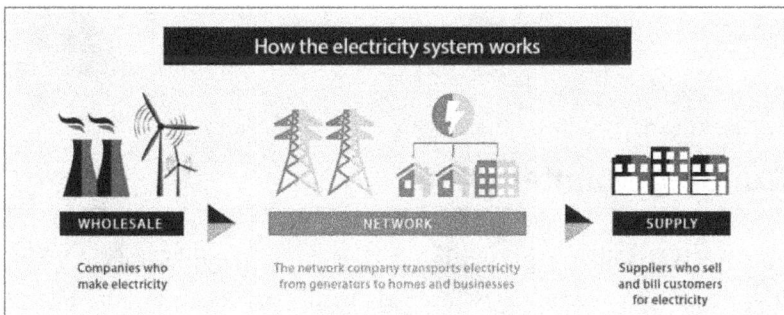

How the electricity system works

WHOLESALE
Companies who make electricity

NETWORK
The network company transports electricity from generators to homes and businesses

SUPPLY
Suppliers who sell and bill customers for electricity

Electricity distribution

(1)	Scottish & Southern Electricity Networks	(5)	national**grid**
(2)	SP ENERGY NETWORKS	(6)	UK Power Networks
(3)	NORTHERN POWERGRID	(7)	Electricity Networks
(4)	electricity north west	(8)	ESB NETWORKS

Meter Point Administration Number - MPAN

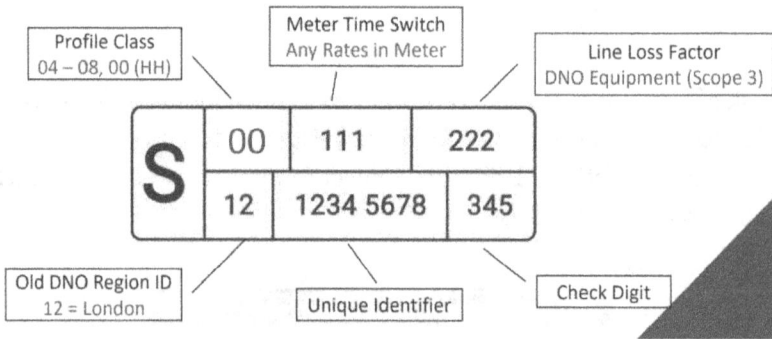

Profile Class 04 – 08, 00 (HH)	Meter Time Switch Any Rates in Meter	Line Loss Factor DNO Equipment (Scope 3)
S 00	111	222
S 12	1234 5678	345

Old DNO Region ID 12 = London — Unique Identifier — Check Digit

Building Maintenance

You're tasked with a crucial role, under the Health and Safety at Work. It's your duty to ensure – where reasonably practicable – the health, safety, and welfare at work. But what about guests in a hotel? Well, you have responsibilities towards them too, under the same Act. Now, let's talk about your workplace. It's essential that certain equipment, devices, and systems are kept in efficient working order – not just to tick a box, but to genuinely safeguard health, safety, and welfare. For instance, think about the mechanical ventilation systems or any equipment and devices that could pose a risk if they malfunctioned. Yes, these need regular maintenance, not only to function well but also to prevent or reduce hazards. Remember, maintenance isn't just about fixing things when they break; it's about keeping everything running smoothly and ensuring a safe environment for everyone. Whether you're managing an office building or a busy hotel, the goal remains the same: a safe, efficient, and legally compliant operation. Examples of equipment and devices which require a scheme of maintenance include:

- Emergency lighting
- Ventilation
- Anchorage points for safety harnesses
- Water monitoring
- Lifts
- Fire extinguishers
- Smoke alarms
- Backflow valves
- Pressure vessels
- Fire Monitoring Systems
- CCTV
- Kitchen Equipment

The condition of the buildings needs to be monitored to ensure that appropriate stability and solidity is provided. Statutory maintenance generally has fixed intervals of time (such as monthly, quarterly, or annually). It is non-discretionary as it responds to health and safety requirements and the consequences of failure of the assets are not considered tolerable. Also, this type of maintenance can be performed by licensed contractors under a service agreement.

What is Compliance Maintenance?

Compliance maintenance refers to the act of ensuring your facility is adhering to all relevant statutory and regulatory laws and legislation.

There are a vast number of pieces of national and international legislation that a business must follow, and failure to do so can result in a heavy fine, and sometimes even court cases. What compliance maintenance services you choose will depend on a number of things, including the size of your property and the equipment that you use.

Fire:

Ensuring that your facility conforms to all fire safety legislation is essential to a safe business. All aspects of fire safety are managed by a 'responsible person' who needs to ensure that all fire systems are inspected and checked regularly in order to comply with relevant British Standards, fire risk assessments are carried out, staff are trained, fire emergency evacuation routes planned and life safety systems in operation.

Electrical:

Any organization must follow the Electricity at Work Regulations which requires that all electrical systems are maintained in a manner that will prevent danger. This includes regular inspections and tests to ensure systems are safe. A detailed preventive maintenance plan can cover all electrical compliance necessities, including running Electrical Installation Condition Reports and Portable Appliance Testing (PAT).

Gas:

Any gas systems present on a building must be checked regularly to ensure that they are safe and suitable for use. This includes safe installation, servicing, inspection and certification for all types of gas equipment. These must be carried out by registered gas-safe engineers. These services can be carried out as part of a compliance maintenance plan, as well as other gas compliance services, such as energy surveys and efficiency assessments.

Water:

Water has the potential to be incredibly harmful as if not checked regularly, it can harbour deadly diseases, such as legionella. All water systems must have a valid Legionella Risk Assessment to ensure that the risk of legionella is reduced. Many businesses are also required to carry out monthly testing of water temperatures to check that water is kept at an optimum temperature to avoid any water-borne diseases.

Asbestos:

Unfortunately, asbestos in buildings is still an issue, with many buildings still having the toxic substance. For that reason, The Control of Asbestos Regulations 2012 was put in place, requiring all building owners to identify and safely remove any materials that may contain asbestos. Sometimes removal is required, although often it just needs to be managed effectively. Staff working in a building with asbestos also must be trained to deal with it and notified of its presence. The removal of the substance must be carried out by licensed asbestos removal professionals.

Air Conditioning:

Under the Energy Performance of Buildings Regulations (2007), air conditioning systems must undertake energy inspections at intervals of no more than five years.

F-gas regulations require inspection and certification:

The regularity of these depends on the weight and number of units:

Every 12 months – buildings over 3kg refrigerant, usually 1-15 air conditioning units

Every six months – buildings over 30kg refrigerant, usually 15-75 air conditioning units

Every three months – buildings over 300kg refrigerant, usually more than 75 air conditioning units

The inspections must be carried out by an approved inspector, and should take into account design, installation and operation of the system.

To aid you in meeting these compliance obligations, the Chartered Institution of Building Services Engineers (CIBSE) can be an invaluable resource. CIBSE is an international professional engineering association representing building services engineers. These professionals are known by various titles including mechanical and electrical engineers, architectural engineers, and technical building services engineers, among others. CIBSE is a recognized member of the Construction Industry Council and is regularly consulted by the government on matters relating to construction, engineering, and sustainability. Additionally, it's licensed by the Engineering Council to assess candidates for inclusion on its Register of Professional Engineers. Turning to CIBSE for guidance can help you navigate the complexities of compliance in building maintenance, ensuring you're not just compliant, but also contributing to a safer and more sustainable future.

Maintenance of buildings and engineering services requires a dedicated workforce with the necessary skills and expertise. This may be direct labour or a contracted service, or a combination of the two. In any property there should be a maintenance programme identifying what tasks are undertaken and to what frequency, and a detailed register listing all the PPM status. PPM stands for Planned Preventive Maintenance, which is also called planned maintenance or scheduled maintenance. PPM maintenance is carried out on an asset (like a piece of equipment, a property, or an element of a property) on a regular basis. A good PPM system is performed to help preserve the properties' condition and prevent problems from occurring. It's a proactive approach to maintenance, designed to avoid failures, breakages and unexpected maintenance costs or unplanned disruptions from reactive works. The biggest benefit of planned maintenance is that it ensures any potential issues are identified and addressed before they develop further, minimising any repair costs incurred. Typically, a well-implemented planned maintenance schedule will reduce reactive maintenance costs by 12-18%. In other words, making a Planned Preventive Maintenance schedule is a must-have rather than a nice-to-have.

Significant benefits include:

- Ensures health and safety compliance
- Less need for major unplanned repair work, through regular maintenance inspections, you are less likely going to be ambushed by unplanned works.
- More efficient use of workforce and budget, maintenance work and financial costs can be spread more evenly throughout the year
- Meeting warranty requirements
- Increases value to preserve and enhance property assets, a property in poor shape is, of course, unlikely to get a good price

Often PPMs are misinterpreted as a superfluous cost; however, without preventive maintenance, the assets risk higher costs for loss of business due to unexpected equipment breakdown. Hotels could risk losing thousands of rental incomes due to an uninhabitable condition caused by poor condition. It is now clear that maintenance can be split into two main categories. The reactive maintenance strategy which allows only for making repairs following a failure, generally more costly and urgent; and the planned preventive maintenance strategy which we have just mentioned where the maintenance tasks are scheduled ahead of time. It's crucial to pick a maintenance strategy that supports both effective asset management and facility management. When repairs are reactive and unplanned, it can really hit your bottom line hard. A well-crafted Planned Preventive Maintenance strategy can significantly reduce the risks associated with property investment. Without careful budgeting, being responsible for a property can become expensive.

Here are some key topics you should consider for an effective PPM:

1. Understand the property use: Get to grips with how your property is used to prevent recurring issues.
2. Assess the building: Conduct a thorough assessment of the building's current condition.
3. Craft a custom strategy: Create a tailored strategy that keeps the building in good shape and plans for future improvements.
4. Report regularly: Produce detailed reports that track progress and issues.
5. Get to know the warranties and insurance: Familiarise yourself with the building's warranties and insurance requirements.
6. Manage your assets: Take charge of both property and asset management.
7. Budget annually: Generate a comprehensive annual budget.

Now, you might be thinking of using Excel spreadsheets and calendar reminders to manage your PPM schedule, and that's a great start. However, if you're managing multiple properties, the workload can become overwhelming pretty quickly. Tasks might slip through the cracks, and suddenly, you're faced with a mountain of reactive maintenance bills. This is where Computer-Aided Facility Management (CAFM) technology comes into play. It ensures that every planned and cyclical task is allocated and tracked from start to finish. Modern management software can be a game changer, helping you stay on top of everything that needs attention, including daily tracking of all remedial work. Whether you're coordinating planned tasks via an internal team, external contractors, or a mix of both, having a system in place is essential.

Features like automated reminders can also keep these tasks top of the mind for both you and your contractors, making sure nothing gets missed.

Remember, a proactive approach not only saves money in the long run but also keeps your property in peak condition, enhancing its value and appeal.

The criticality of the critical spare part

Let's talk about an essential aspect of building management that might not always grab the headlines but is fundamental to keeping a building operational: managing critical spare parts. This isn't just about having a few extra nuts and bolts lying around; it's about understanding and maintaining the backbone of your building's functionality. Critical spare parts are the components vital to your building's day-to-day operations. These include crucial elements of your HVAC system, elevator mechanics, plumbing essentials, and fire safety equipment. The failure or absence of any of these parts can lead to operational disruptions, safety risks, and unexpected expenses. The main goal of stocking these parts is to ensure your building's operations never skip a beat. When equipment breaks down, having the necessary spare parts on hand means you can quickly address the issue, minimizing any impact on occupants and operations. Without these parts, you risk prolonged downtime and the complications that come with it. Having a well-managed inventory of critical spare parts can lead to significant cost savings. Procuring parts on an emergency basis often comes with premium prices and additional shipping costs. By maintaining a strategic stock of essential parts, you can avoid these extra expenses and ensure you're prepared for any situation. Many critical spare parts are directly related to safety systems, such as fire sensors. The immediate availability of these parts is non-negotiable for ensuring the safety and well-being of building occupants. Regular maintenance and prompt replacement of these components are crucial for upholding safety standards. Routine maintenance is key to extending the lifespan of your building's equipment. This includes replacing parts before they fail. A reliable inventory of spare parts ensures that maintenance schedules are adhered to, preventing minor issues from escalating into major problems. In the long run, this approach not only saves money but also contributes to the sustainability of your building's operations.

The Art of Managing Critical Spare Parts

Understanding the importance of spare parts is one thing; managing them effectively is quite another. Here I suggest some straightforward steps, which may help you to keep your spare parts organized and ensure your building keeps running smoothly.

Inventory Management: Keep a current list of all your spare parts. Make sure this list includes what each part does, how many you have, where they're stored, and any other important details. This helps you know exactly what you have and where it is.

Demand Forecasting: Look back at your building's repair and maintenance history to predict which parts you might need soon. This way, you can make sure you have the right parts on hand when you need them without keeping too much stock.

Supplier Relations: Build good relationships with your suppliers to make sure you can get spare parts quickly and without spending too much money. This is especially important for parts that are hard to find or special in some way.

Regular Audits: Check your spare parts stock regularly. This helps you keep an eye on what you have and find out if there are any parts you don't need any more or have too many of.

Lesson learned.

I want to share a personal experience that taught me a valuable lesson about being prepared and understanding the importance of having critical spare parts on hand. This story revolves around a major hiccup I faced with the A/C system, leading to some hard-earned insights. One day, the A/C system for an entire floor went down due to a severe compressor failure. The twist? The specific model we needed wasn't available locally. We ended up having to order a replacement from our supplier in another country. With international shipping and all the logistics involved, it took a full two weeks to get the new compressor delivered and installed. The result from this breakdown was significant and highlighted several areas we needed to improve. Every room on the affected floor was out of order because the temperatures were just too uncomfortable. Since these rooms made up a large chunk of our total capacity, the hit to our potential booking revenue was massive. But it wasn't just about the money. The service disruptions were a big deal too. We had to cancel bookings for those rooms and move some guests to other hotels, which wasn't a great look for us. It caused quite a bit of guest dissatisfaction and potentially harmed our reputation for service excellence. The indirect costs, including those guest relocations, added to the financial sting. Having to wait two weeks for the compressor, combined with the revenue loss from not being able to book those rooms, turned into a significant financial setback. This whole ordeal was a stark reminder of how crucial it is to have a supply of critical spare parts ready to go, especially for specific equipment.

Team Management

As a leader or future leader, you aim to build the skills and capabilities of your teams, enabling them to take responsibility and make decisions that support the smooth running of the business. The goal is to develop and grow their confidence so they can make these decisions effectively. We strive to cultivate a clear and recognizable culture within the business that empowers the team to grow in their roles; also, the guests expect swift, positive service from everyone, therefore the team's confidence is crucial. Try to list ways to empower your team, identifying any training or coaching events that would facilitate achieving great assuredness and ask yourself how you can build the confidence of your team, so they feel capable of taking on greater decision-making responsibilities. The generational gap in the workplace is another key factor to consider. Each generation grows up in a different context and, as a result, may have different work expectations. For instance, Gen Z is heavily tech-reliant and comfortable using social media platforms, while older generations may prefer other forms of communication. Challenges in managing generational gaps in the workplace can arise from misunderstandings. Each generation can have its own preferences and expectations when it comes to completing job responsibilities. For example, Gen X and baby boomers may be more deferential to authority than Millennials or Gen Z. Additionally, each generation can have a different preferred communication method, making a well-thought-out communications strategy essential to mitigate the generational gap.

Design Your Communication Strategy

Different generations receive information in various ways; therefore, managers should tailor their communication methods to each generation's strengths, personality, and aspirations. Mobile and desktop email appeal to Gen X. You can capture their attention through training content as well. In fact, more than ten percent of them have completed a doctorate or other professional degree, which should give you an idea about how to communicate with them. For Millennials, it's all about texting as it appears to be their most preferred communication method. They find it easy and common. Generation Z is the generation of the smartphone. To them, email seems to be old school. They are known for keeping it brief, meaning they expect most communication to fit within the context of their phone screen.

Listen

As simple as it might seem, you can learn a lot by just listening to your team. If you want an open communication between generations and build relationships, ask them about their preferences, interests, and expectations, and then listen carefully. Engage in a dialogue and see where it leads. Leaders who listen to their team members from the time they join the company understand that high-performance cultures are mainly built on relationships and based on communities of common interests.

Computer Aided Facilities Management

Imagine yourself as a professional, equipped with the power of a tool that transforms the way you handle your daily responsibilities. When we synergise CAFM with other cutting-edge technologies such as AI or IoT, it's not just about reducing operating costs; you are also stimulating your organization's top-line growth. Imagine dealing with old, disconnected systems like shared drives, and spreadsheets to manage your facilities. Feels chaotic, doesn't it? Now, contrast that with the streamlined approach offered by a Computer Aided Facility Management (CAFM) system. It's easy to see why more and more organizations are choosing to adopt the CAFM. Of course, every organization has its own unique needs. But, regardless of these differences, there are some benefits that just about everyone enjoys when they install a CAFM system. How exactly does this happen? An effective CAFM solution should offer an intelligent, helpful user interface designed to support your strategic decisions and simplify the management of your facilities. This can come in the form of interactive databases, interactive graphics and data management displays, but without the ability to digest and understand this information, you're missing out on the true power of CAFM. A top-class Computer-Aided Facility Management system can truly transform how you manage your assets. Imagine having the ability to track every chair and desk along with crucial maintenance data – no more gaps in your asset knowledge. You'll enjoy full control over costs and resources, which is a game-changer for any business. With planned preventive maintenance, you'll gain a better understanding of your assets and their specific needs. This allows you to schedule maintenance tasks well in advance, saving time and significantly boosting your business's operational efficiency. Stock control is another area where a CAFM excels. Know your stock levels in real-time and plan ahead when supplies are running low. This foresight helps cut down on overspending and positively impacts your budget. When it comes to technical assistance, the help desk feature of a CAFM system enables you to balance proactive and reactive management efficiently. Get real-time updates and log issues as they arise, ensuring nothing slips through the cracks. Space management is also enhanced with CAFM. By incorporating floor plans and layouts, you can optimize how these spaces are managed and occupied. Track which spaces are most frequently used and make informed decisions about underutilised areas. Building operations benefit too, with remote management capabilities that allow you to control various aspects of your buildings right from your desk. This improves efficiency and reduces your daily workload. CAFM also supports your team by providing dedicated workflows and risk reports, promoting a people-focused working environment. It ensures that everyone is equipped with the tools they need to succeed. Lastly, contractor management is streamlined with CAFM, ensuring your contractors are compliant and well-managed. This system provides checks and balances to keep work on schedule and maintain high standards of compliance and efficiency. I believe that CAFM systems will be commonly used in the near future from the majority of the organizations as you're not just maintaining your facility; you're enhancing its value and operational capacity, ensuring everything runs as smoothly and efficiently as possible.

Request for Capex

I think that CAFM is more than a software package; it's a game-changer for facilities management and probably many others would think the same… so… Why wait, let's install it!! Well, that could be a bit of a problem as it can be one of the cases where the process of having a green light to proceed with the investment can be quite challenging as explaining the benefits to your C-suite or directors for a new system or introduce a very technical project can be difficult and sometimes we can miss the real goal. I'm sure many of you have had this issue at least once! We need to keep in mind that one of our main tasks is to bring complex stuff to the table and translate them as simply as possible, allowing non-technical people to understand them. It can be ok on a daily basis or for low/medium investments, but when the big money is needed, that's a different story, as thousands of questions are asked, more details need to be provided and pros and cons studied in depth.

If you want to win the argument for funding, you've got to build a business case. You've got to demonstrate that your purchase is required – that it will not just be a cost, but will actively cut costs and improve performance and enhance profitability. Who do you need to persuade to get the project approved? Usually, the Board of Directors and finance team are the two main key players. Within the realm of finance, where accuracy and figures reign supreme, the mission to enhance budgeting precision and spending visibility takes centre stage.

Let's say you want implement the CAFM system in your building, it's essential to understand first the project strengths, second, understand what the directors want to hear, and third, use the project strengths as bullets to show how your proposed solution can support the company. For example, in this case the CAFM can empower you to take control and naturally report on budget versus spend, providing the finance team with the tangible evidence they seek, or the ability to unlock greater value for money from suppliers, satisfying the finance team's quest for efficiency and fiscal optimization. The BODs, with their financial judgement and strategic foresight, are driven by a twin-pronged objective. Firstly, they have a keen interest in ensuring that teams operate like a well-oiled machine, delivering tip-top results that reinforce the company's performance. However, their eyes never stray far from the bottom line, where the language of profit and investment returns do the talking. A robust vision of the company's profitability and the dividends their investment will produce is integral to their decision-making process in procurement. There are few common goals that the directors and the finance team share and can be grouped as follows:

Save Time – The company aims to not just save valuable time but also to make its operations smoother. They are always on the hunt for ways to work smarter, not harder, aiming to enhance efficiency and productivity in every task they undertake.

Minimise Risk – The organization wants to minimize the chance of business errors and non-compliance issues. Everyone needs to pull together to safeguard vital data, protect income streams, and uphold the company's reputation. They aim to steer clear of any fines or possible legal problems that could arise from poor management.

Save Money – Every team in the company shares a collective aim to boost profits. They strive to cut down on unnecessary costs and squeeze every penny of savings they can.

Enhance Service – Using information wisely, automating tasks, and working better together, both the facility management teams and suppliers can keep a closer eye on their Service Level Agreement (SLA) performance. This transparency allows them to consistently deliver top-notch service.

Support Strategy – They require tools that align with the company's strategic goals. These include breaking down internal barriers, embracing digital change, striving for constant progress, and unlocking the potential for future business expansion.

Now that we know what the directors are interested in, we can work around it, highlighting your project strengths. I find presentations very helpful to support the case and keep the discussion on track helping to transform "numbers on a page" into something more tangible. But you don't need a master how to write a budget winning proposal – you just need to write what you know. Below are some tips

Emphasise clarity and brevity, ensure your writing is lucid and concise, highlighting exactly what the business will gain. Avoid unnecessary topics and keep your message succinct.

Convincing Story – create a coherent and concise narrative in your proposal:

- explain the problem,
- potential solution,
- potential risks of inaction
- your vision of a more efficient future.

Use testimonials and real case-studies to support your project, address your audience's priorities, focusing on the real-world implications of your solution, its response to the business issues, and its full benefits.

Insert visuals, boost your document with charts, tables, and screenshots, especially images of your proposal.

Show potential savings and above all, quantify the potential savings your solution could produce. Seek assistance from your provider/partner to access and present this crucial information realistically.

Great! We have a deep understanding of our project benefits (CAFM in this case, but it's the same principle for any request for funds). We now know what the directors are interested in and how to approach them to proceed with your project. All three elements which can increase the probability of success when we ask for an RFC.

Inspections

Inspections are a critical component of any maintenance programme designed to protect your assets at any level. They help you check the condition of your equipment and determine what tools, materials, and labour will be needed for maintenance. Essentially, a maintenance inspection is your way of assessing the condition of your property. It's a proactive step that helps you catch minor issues before they blow up into costly repairs. Ideally, these inspections should be scheduled, completed on time, and meticulously documented using a maintenance management system or similar software. This approach not only keeps things organised but also ensures that nothing slips through the cracks.

There are several types of maintenance inspections to be aware of: preventive, predictive, condition-based, corrective, and unscheduled.

Each type serves a unique purpose, from routinely preventing deterioration to predicting future issues based on current data. Integrating these inspections into your maintenance strategy helps you maintain a robust defence against unexpected failures and costly downtime.

Preventive Inspection is a process where the quality is inspected to ensure that standards are met. A preventive inspection aims to find potential problems before they arise to minimise risk and improve efficiency.

Predictive Inspection is a process that uses sensors to collect data about an object, structure, or material to predict the probability of failure. Predictive inspections are designed to identify potential problems before they become unsafe or expensive ones. The information collected can be used for prognosis, maintenance planning, and even forecasting the risk of failure.

Condition-Based Inspection is a process in which the inspector examines the item to determine if it needs repair. It can be used in various settings, such as in maintenance departments for equipment that requires periodic check-ups or by homeowners doing their own inspections before listing or selling their houses. This inspection differs from a risk-based inspection, which determines if something poses an imminent safety concern.

Corrective Inspections help to identify and fix any issues that may have been missed in the inspection process. A corrective inspection is a type of inspection that focuses mainly on problems identified by the guests.

Unscheduled Inspection is a process in which an inspector visits a site without warning to make sure that the standards are met constantly.

Non-Destructive Inspection (NDI) is a process that detects mistakes or defects in materials, repairs, or other structures without causing any damage. These inspections are used to identify problems that cannot be seen by the naked eye.

Furthermore, a good maintenance inspection programme should include:

Safety and Risk management inspections: These inspections can include everything from checking and restocking first aid kits to assess if the items stored in a specific place can cause hazards or fire risks. It can also include checking those components of systems and equipment that are critical for safe operations.

Failure finding inspections: These inspections check the operation of back-up or protective devices that cannot be readily checked unless a primary system fails. Failure instances must be simulated to test these components of equipment or machinery.

Lighting inspections: Bulbs and emergency lights should be checked regularly and replaced in a group when they begin to fail for efficiency. In addition, a technician should examine controls, transformer systems as well as cables, hardware, and gaskets on exterior lights.

Electrical inspections: From simple battery replacement such as flashlights or testing electrical equipment (PAT testing).

HVAC Inspections: Air intake, filters, motors, and ductwork should be inspected and cleaned regularly. It is important to check drainage functioning of condensation pans and secure loose panels, guards, and hardware.

Building interior inspections: Check walls, ceilings, and floors for damage, leaks, or other deterioration. Be sure to remove hazards and ensure doors and locks work correctly. Toilet areas should be inspected, and DDA alarms should be tested.

Building exterior inspections: Check paint, walls, windows, and doors regularly as well as any foliage that may damage the walls or foundation. Inspect the roof, drains, and gutters.

Inspect sidewalks, driveways, and railings for hazards and damage.

Plumbing inspection: The flow circuits should be checked at least once a year for leaks, noises, and damage. Water boosters, water chillers, condenser fans, and circulation pumps must be lubricated, and water boilers and heaters should be fire-tested. Sewage and sump pumps should be inspected regularly and replaced as needed.

Increased safety and compliance aren't the only benefits of performing inspections. If inspections are regularly done, you can lower maintenance costs, extend the life of equipment, and increase operational productivity. All equipment wears, but that doesn't mean it has to break. When you can determine how often a particular type of equipment needs a certain repair, you can schedule that repair before failure occurs, thereby reducing repair costs and minimizing downtime. It's essential to identify recurring problems so that their causes can be corrected. Inspections give you the data to quantify trends issues, as well as to isolate causes. Often, recurring issues are caused by systematic misuse, and inspections can bring this operational issue to light. When inspections are planned the personnel will take better care of the property because regular inspections hold them accountable; in fact, often simply doing pre-use inspections causes employees to have a higher respect for the building condition, making the threat of consequences largely unnecessary. A maintenance inspection checklist should be used to record the equipment and services that need to be looked at during the scheduled time. It ensures that everything is in working condition and there are no safety hazards. A Computerised Maintenance Management System (CMMS) can help to increase productivity during the process.

A maintenance inspection completed means that the remedial works have been accomplished (proof of work) and the necessary actions have been executed.

Negotiation and tendering process

Being a leader often means managing a significant number of firms to ensure everything runs smoothly. One of your key responsibilities is choosing the right partners to collaborate with, aiming for the most successful outcome for your business. This is where mastering negotiation becomes invaluable. Negotiation is a powerful skill – it's the art of discussion aimed at reaching an agreement where everyone feels they've gained something. In these discussions, different parties come together to hash out their varying needs and goals, aiming to find a solution that everyone can agree on. This requires a bit of give and take, and your goal should be to foster a courteous and constructive interaction that ends in a win-win situation. Ideally, a successful negotiation is one where you can afford to make concessions that don't mean much to you but mean a lot to the other party. The best negotiations leave each party satisfied and eager to do business together again. It's not just about striking a deal – it's about building relationships that pave the way for future cooperation.

Good negotiations contribute significantly to business success, as they:

1. help you build better relationships
2. deliver lasting, quality solutions
3. help you avoid future problems and conflicts.

To get the best outcomes, it's necessary to understand the steps involved in the negotiation process. While many negotiations are straightforward, some will be among the hardest challenges you face. Your success will depend on planning and preparation. Always approach negotiations with a clear set of strategies, messages and tactics that can guide you from planning to closing. As a negotiator you must master written, verbal, and non-verbal communication, and adopt a conscious, assertive approach.

Good negotiators are:

•   flexible

- creative
- aware of themselves and others
- good planners
- honest
- win-win oriented
- good communicators.

During a negotiation, the communication style you choose can usually be passive, aggressive, or assertive and can significantly influence the outcome. Opting for an assertive style generally increases your chances of securing successful outcomes for your business.

Let's break it down:

Passive communicators often use vague language, exhibit under-confident body language, and tend to give in to requests too easily. While this might seem non-confrontational, it often doesn't lead to the best outcomes for your interests.

Aggressive communicators take a confrontational stance, which can alienate other parties and potentially derail the negotiation process. Although it might feel powerful in the moment, this approach can burn bridges and harm long-term relationships.

Assertive communicators, on the other hand, strike a balance between confidence and consideration. They maintain a successful dialogue and are more likely to facilitate beneficial outcomes. To communicate assertively, adopt a strong, steady tone of voice and stick to the facts rather than getting emotional or critical. Sharing your opinions clearly and starting sentences with "I" (as opposed to launching direct criticisms with "you") can help create a more collaborative and less confrontational atmosphere. Embracing assertiveness in your negotiations not only helps in reaching agreeable solutions but also builds a foundation for respectful, productive relationships in the business world.

Don't:
- confuse negotiation with confrontation – you should remain calm, professional, and patient

- become emotional – remember to stick to the subject, don't make it personal, and avoid becoming angry, hostile, or frustrated
- blame the other party if you can't achieve your desired outcome.

Do:
- be clear about what you need from the other party
- be prepared, think about what the other party needs from the deal, and take a comprehensive view of the situation
- be consistent with how you present your goals, expectations, and objectives
- set guidelines for the discussion and ensure that you and the other party stick to them throughout the entire process
- use effective communication skills including positive body language
- prepare for compromise
- strive for mutually beneficial solutions
- consider whether you should seek legal advice
- ask plenty of questions
- pay attention to detail
- put things in writing

Even with the best preparation, you may not always be able to negotiate a successful outcome. You should plan for what to do in case negotiations fail. If you allocate time and resources to planning alternative solutions, you can avoid unnecessary stress and poor business outcomes. Having an alternative plan will help you to reduce your own internal pressures, minimise your chances of accepting an offer that is not in your best business interests, and set realistic goals and expectations. One of the tools which can be used to start the negotiation and facilitate the process is tendering. Tendering is a formal process where companies are invited to bid for contracts from public or private sector organisations, which need specific skills for a project, or goods and services. The three types of tenders are:

Negotiated tenders, open tenders and selective tenders. These differ in terms of who can bid for the tender, as well as their advantages.

Negotiated tender

The negotiated tender involves reaching out to a single supplier that is highly suited to a specialist contract or reaching out to a supplier to extend an existing contract. These types of contracts are prevalent within the construction and engineering industry. The supplier and tenderer are acquainted with each other and their work; this can benefit the tendering process by decreasing the duration of the exchange. Additionally, there's quality assurance and a higher expected success rate given the relationship between the tenderer and supplier. The downside is that the price of the contract is likely very expensive and sometimes the negotiation isn't as simple as you may think. In this case the absence of competition may result in a prolonged negotiation period to ensure both parties receive a fair deal.

Open tender

This is the main tendering process used today. This process invites everyone to submit a tender proposal, offering an equal opportunity to all. This is believed to stimulate competition, ensure that everyone gets a fair chance, and allows for the most cost-effective option. There are certain disadvantages with open tendering. It's known to be slow and costly, as it garners the attention of many suppliers, many of which are unsuitable but still have to be acknowledged. This may result in a lot of time, money, and effort wasted in trying to submit and analyse a tender proposal. Tenderers can minimize these downsides by requiring bidders to go through a pre-qualification process; these may help reduce the bidding firms dramatically. Lastly, although competition is beneficial to the tenderers in terms of competitive price, they have to be careful because while under-priced tenders may seem appealing, there might be a price-quality trade-off at play. A lower price may involve low-quality material, low quality of work, and other such factors to cut corners. Open tenders can either be single-stage or two-stage. Single-stage tendering occurs when there's enough information to come up with a price on the spot. A two-stage tendering, on the other hand, means that information is insufficient, so two separate agreements occur. The first one settles technical specifications, and the second one settles the price.

Selective tender

Pre-selected number of suppliers are invited to bid on the contract. This type of procedure is best suited for complex contracts that only align with a few firms' capabilities and aims to improve the quality of bids received. These firms generally have a track record that signifies competence for similar projects in terms of complexity, size, and nature. Generally, no more than four suppliers are invited to submit a tender. The benefits of pursuing selective tendering are that it results in a quicker conclusion than the open tendering and could be more efficient as there should be lower documentation cost. There are, however, certain disadvantages as selective tendering tends to exclude smaller or new suppliers who are in the process of establishing themselves in the market (often smaller companies are more customer service oriented showing positive attitudes and performances, demonstrating an awareness and willingness to respond to customers in order to meet their needs, requirements and expectations). Similar to the open tendering, selective can be either single-stage or two-stage. It's important to use correct type of tendering procedures you require and understand how it can impact the tendering process. Ultimately, when you are closing the negotiation just take a moment to revisit your objectives.

Once you feel you are approaching an outcome that is acceptable to you:

- look for closing signals
- fading counterarguments
- tired body language from the other party
- negotiating positions converging
- articulate agreements and concessions
- make 'closing' statements
- 'That suggestion might work.'
- 'Right. What's the next step?'
- Let's make it happen!
- get agreements in writing as soon as you can
- follow up promptly on any commitments

Section
4

Efficiency

Think of energy as the ultimate multitasker; it can transform in so many ways – into motion, light, heat, you name it. But here's the cool part: it's always conserved. That means we can't create new energy or make it vanish into thin air. It just changes from one form to another. It's like that saying, "You can't get something for nothing." Energy always must come from somewhere, and it's not going anywhere – it just keeps morphing into different forms.

Let's have a look at the units of energy:

- Kilowatt hour (kWh): commonly used in the electricity supply industry
- British thermal unit (Btu): the old imperial unit of energy, it is still very much in use.
- Therme: a unit that originated in the gas supply industry. It is equivalent to 100,000 Btu.
- Tonne of oil equivalent (toe): a unit of energy used in the oil industry.
- Barrel: another unit of energy used in the oil industry.
- Calorie: commonly used unit of energy in the food industry. It is in fact the amount of heat energy required to raise 1 gram of water through 1 °C.

You have probably heard the push about going green and saving energy. The green movement seems to have taken over the world, empowering eco-friendly people every day. There are a number of reasons why you should consider cutting back on energy consumption. First of all, reducing energy usage limits the number of carbon emissions in the environment. Carbon emissions play a significant role in climate change, which is thought to be the cause of powerful natural disasters in recent years. With billions of harmful emissions in the atmosphere, cutting back is always a good thing. In turn, conserving energy produces a higher quality of life. Reduced emissions result in cleaner air quality and it helps create a healthier planet, or at least helps sustain the resources we already have. Being conservative with energy can ensure that lakes, trees and animals are around for future generations. Using energy more efficiently is one of the fastest, most cost-effective ways to save money, reduce greenhouse gas emissions, create jobs, and meet growing energy demand. The benefits of energy efficiency include:

Environmental: Increased efficiency can lower greenhouse gas (GHG) emissions and other pollutants, as well as decrease water use.

Economic: Improving energy efficiency can lower your property utility bills

Utility System Benefits: Energy efficiency can provide long-term benefits by lowering overall electricity demand, therefore reducing the need to invest in new electricity generation and transmission infrastructure.

Risk Management: Energy efficiency also helps diversify utility resource portfolios and can be a hedge against uncertainty associated with fluctuating fuel prices.

Imagine this: even the smallest day-to-day actions can pack a punch in the fight against energy waste. It's vital for any business – whether it's a buzzing hotel or an office building – to embrace a plan that rallies the whole team around saving energy. It's all about nurturing an energy-saving culture, where every little action counts toward a bigger goal of sustainability. So, what can we do? Starting with simple things. Turning off workstations after the day's work – computers, phones, chargers, the works. It's also smart to think about when we're using gas and electrical appliances. Do we need them on all the time? Cutting back on usage means we trim down on both costs and our environmental footprint. It's a win-win. Water is another big one. Making sure those taps are firmly off to stop any wasteful dripping is key. And let's talk about air conditioning. Sure, we want to stay cool, but let's not go overboard. Keeping doors and windows closed keeps the cool in and the heat out, helping us use less energy to stay comfortable. Every one of these little habits adds up to a pretty significant impact. It's the small gestures that can steer us towards a more sustainable future.

Many energy efficiency measures focus on mechanical and electrical equipment. Energy performance, and architectural decisions can impact both lighting requirements and ventilation loads. Together, ventilation, heating and lighting can account for more than 85% of energy expenses. For this reason, any energy-efficiency decisions on which systems or plants should be installed to replace obsolete equipment will be beneficial in terms of costs, for both short and long term.

Below are listed some of the most common energy-efficient systems:

Use of energy efficient LED lighting; this increases the life of the lighting system, reducing the need for purchasing new bulbs so often.

The use of motion sensors helps to keep energy usage to a minimum by automatically switching lighting off after a pre-set time is reached without movement in the area.

The Voltage Optimization is an efficient transformer with smart controls that dynamically select the optimal voltage within a building to reduce consumption, while continuing to meet building requirements.

Air handling units equipped with heat recovery wheels, a very efficient system as it utilises heat from the extracted air and transfers this into fresh air, reducing the amount of heating needed to warm up intake air.

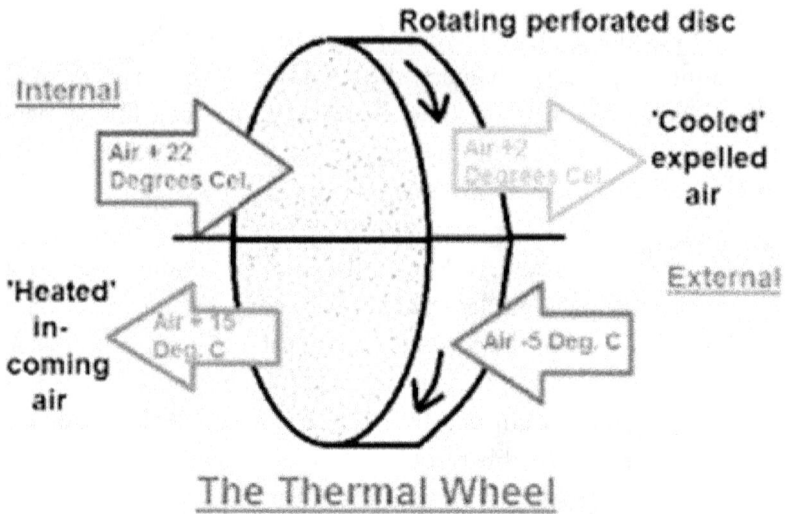

Rotating perforated disc

Internal

Air + 22 Degrees Cel.

Air +2 Degrees Cel.

'Cooled' expelled air

'Heated' in-coming air

Air + 15 Deg. C

Air -5 Deg. C

External

The Thermal Wheel

Geothermal systems are very energy efficient systems. They use the consistent temperature of the earth's surface to heat and cool your home, rather than propane or fuel oil.

Intelligent Demand Controlled Ventilation System for commercial kitchens which adjusts fan speed during times where there is a low usage of equipment.

High-performance heat pumps deliver impressive environmental and energy system benefits. More sustainable to operate than oil and gas boilers, heat pumps can dramatically cut carbon emissions compared to traditional fossil-fuel based heating systems.

Rainwater harvesting systems collect and capture rainwater which runs off large surfaces such as roofs and stores it in tanks which can be underground, such as rain-traps and aqua-banks, or above ground harvesting tanks like water butts in gardens or larger tanks around farm buildings. Rainwater harvesting systems can be used for irrigation and toilets.

RAINWATER HARVESTING

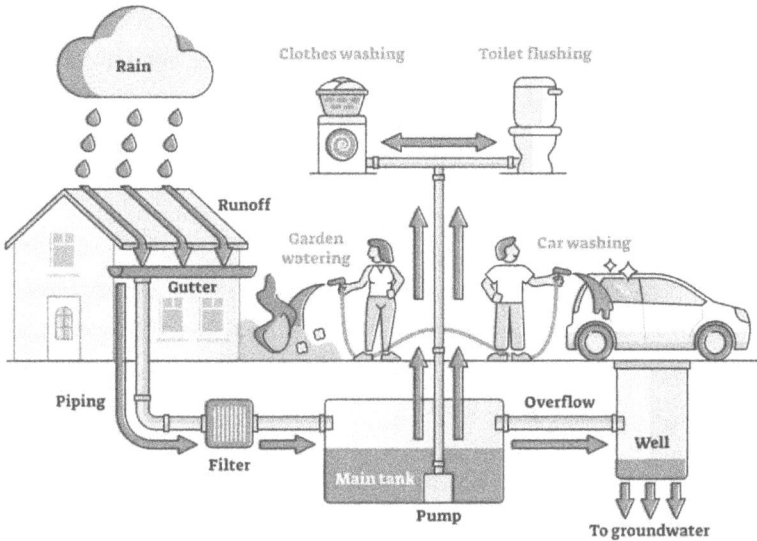

Now, let's talk about power factor – it's an essential aspect of your building's efficiency and definitely worth getting into. Think of it as a scorecard for how well your electrical system is using the power it draws. The closer this score is to one (or 100%), the more effectively you're using that energy. But if it starts to dip, that's like your building's power is getting side-tracked before doing any useful work, and that's not what we want. A low power factor signals means that you're not just losing energy; you're also losing money, and potentially straining your electrical systems. Plus, it's not just about cost. A poor power factor can also increase the carbon footprint of your building, because it requires more energy generation to meet your actual usage needs. So, paying attention to the power factor isn't just about technical efficiency; it's about environmental responsibility and economic sense, too. It's part of a bigger conversation on energy efficiency and sustainability. When you improve your power factor, you're not only optimizing the power you draw from the grid but also minimizing waste – which is good for the planet and your organization's pocket. The power factor (PF) is the ratio of working power, measured in kilowatts (kW), to apparent power, measured in kilovolt amperes (kVA). Apparent power, also known as demand, is the measure of the amount of power used to run machinery and equipment during a certain period. It is found by multiplying V x A. The result is expressed as kVA units. All motors which come in the form of a machine, such as conveyors, mixers, compressors, lifts and escalators, all have an efficiency rating known as a Power Factor. PF has values fluctuating from zero to one, where one is 100% efficient. To better understand, a 96% power factor demonstrates more efficiency than a 75% power factor. Usually, a PF below 90% is considered inefficient. If a circuit were 100% efficient, demand would be equal to the useful power. When demand is greater than the power available, the supply system will be under stress and many suppliers add a demand charge (capacity charge) to the bills to offset differences between supply and demand; that's simply because if demand requirements are irregular, the supplier must have more reserve capacity available than if load requirements remain constant during the time. For most suppliers, demand is calculated based on the average load placed within 15 to 30 minutes.

- The beer is active power (kW) – the useful power is the liquid beer, is the energy that is doing the work. This is the part you want.
- Foam is reactive power (kVAR) – the foam is wasted power or lost power. It's the energy being produced that isn't doing any work, such as the production of heat or vibration.

- The mug is apparent power (kVA) – the mug is the demand power, or the power being delivered by the utility.

Making sense of power factor: The beer analogy

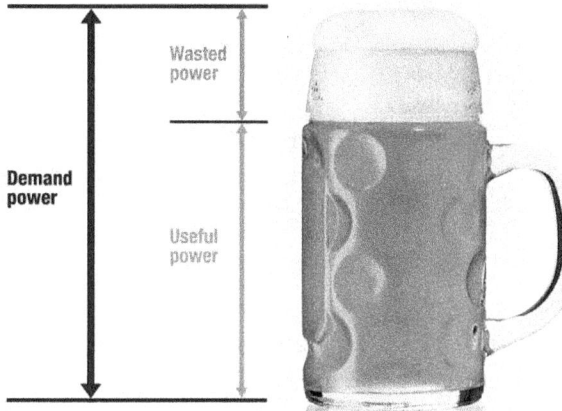

The peak demand is when demand is at its highest and the challenge for suppliers is delivering power to handle every customer's peaks. Low power factor means poor electrical efficiency. The lower the power factor, the higher the apparent power drawn from the distribution network. By installing suitably sized switched capacitors into the power distribution circuit, the power factor is improved and the value becomes closer to one, minimising wasted energy, improving efficiency, liberating more kW from the available supply and saving money. Modern low voltage switchboards are able to calculate the power factor, or in the case that's not available a power quality analyser (see picture below) can do the job; it measures both working power (kW) and apparent power (kVA), calculating the ratio kW/kVA.

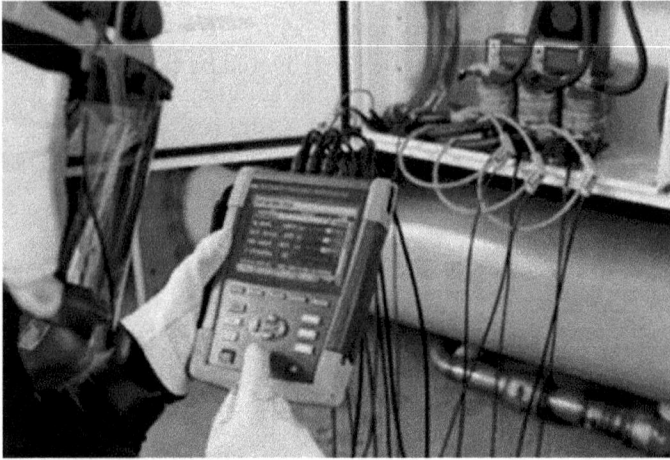

The power factor formula can be expressed in other ways:

PF = (True power) / (Apparent power) OR PF = W/VA

Watts measure the useful power while VA measures supplied power. The ratio of the two is shown below:

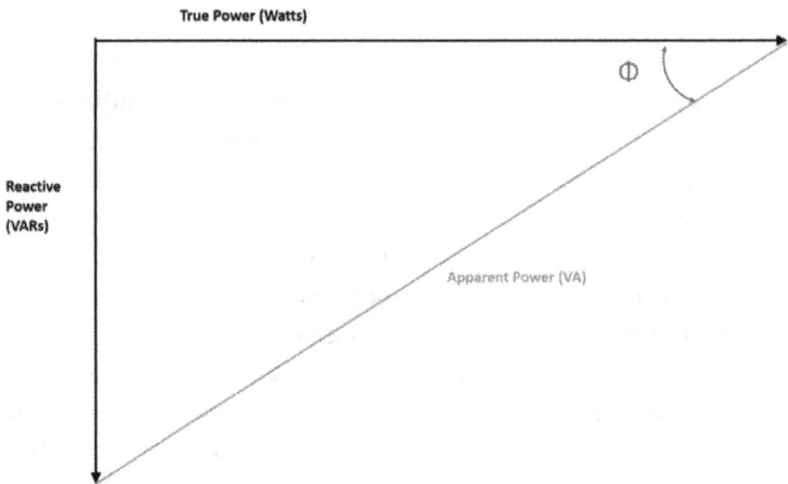

As this diagram demonstrates, power factor associates the real power being consumed to the apparent power, or demand of the load.

As already mentioned, poor power factor means that you're using power inefficiently and this matters because it can result in:

- Heat damage to insulation and other circuit components
- Reduction in the amount of available useful power
- A required increase in conductor and equipment sizes
- It requires a higher current to supply the loads.

Understanding the source and transformation of the power that runs our essential systems is like knowing the roots of a tree. It's the foundation that allows us to grasp how things work and how to manage them better. This knowledge is empowering, quite literally! Let's think about the lifeblood of critical power, essential, and non-essential systems. The power begins its journey as a high-voltage titan. As it navigates through the arteries of the power distribution network, it undergoes a transformation. It's scaled down to become the three-phase and single-phase voltages, a format more suitable for our systems to connect with. It's a fascinating process, isn't it? Let's explore further. There are different categories of AC voltages in the world. Picture them as a sort of power pyramid. At the top, we have High Voltage (HV), then comes Medium Voltage (MV), and finally, at the base, we find Low Voltage (LV). This isn't just a random classification – it's a standardized system defined by the International Electrotechnical Commission. They've precisely sorted these voltages into levels. The IEC are the folks who play a big role in setting international standards for all things electrical, electronic, and related technologies.

So, what does that mean? Well, it establishes a set of nominal voltage ranges for alternating current (AC) and direct current (DC). It's like a rulebook that helps categorize different voltage levels. This provides a uniform language for engineers, manufacturers, and suppliers worldwide, ensuring everyone is on the same page when talking about voltage levels. It's quite a vital tool in the world of electrotechnology. While IEC lays out a set of international standards, it's important to remember that they serve as a guideline. Individual countries can, and often do, establish their own national standards that may vary from the IEC's recommendations. This could be due to a variety of factors, such as legacy infrastructure, local regulations, or specific technical requirements. Hence, while IEC 60038 provides a fantastic starting point for understanding voltage classifications, it's always a good idea to verify the exact standards used in the specific country you're dealing with. This ensures that the electrical systems and components are compatible and safe for use in that particular location. So, while we have global standards, remember to always think locally too! For example, both the UK and the European Union generally follow the same voltage standards as outlined in IEC 60038; in fact, the standard household electricity supply in both the UK and EU fall into the Low Voltage (LV) category. However, it's important to note that while the overarching standards are typically consistent, there can be minor differences in the practical implementation of these standards. For example, the precise voltage can sometimes fluctuate slightly due to factors like the distance from the power source, the time of day, and the demand on the system. Moreover, the type of plugs and sockets used can vary significantly between the UK and many EU countries. The UK uses Type G plugs and sockets, while many EU countries use Type C, E, or F.

Voltage Level	Code	Voltage Rating
Low Voltage	LV	up to 1000V
Medium Voltage	MV	1000V to 35kV
High Voltage	HV	35kV to 230kV
Extra High Voltage	–	above 230kV

Let's investigate why voltage is generated at such a high level. A key reason is to compensate for losses that naturally occur as power travels along the distribution system to reach its final destination (the point of use).

Now, let's touch upon a significant milestone in our power journey. The voltage used throughout Europe (including the UK) has been harmonised; the aim was to establish a common voltage level for the harmonization of electrical equipment and appliances across the European Union. The official harmonization of voltage in the European Union was decided by the European Committee for Electrotechnical Standardization (CENELEC). The decision to standardize the nominal voltage level to 230V for domestic and similar purposes was made in 1987, and the transition period extended from then until 2008 (to find the official documentation and exact dates, you would need to refer to the standards and regulations published by CENELEC or the International Electrotechnical Commission). Prior to this, there were differences in the nominal voltages across countries. For example, the UK had a standard nominal voltage of 240V, while many other European countries used 220V. The harmonized voltage level was set at 230V (with a range of plus ten percent and minus six percent) for single-phase connections (which most domestic appliances use). This means that the voltage can technically range between 216V and 253V, which covers the old 220V and 240V standards. This change was largely administrative. In practice, the actual voltages supplied have not changed significantly; older appliances that were designed for 220V or 240V can still operate under the harmonized system. It's important to note that three-phase voltages were also harmonized at the same time, set at a nominal 400V. This harmonization greatly simplified the manufacturing and distribution of electrical goods across Europe, as they could all be designed for the same nominal voltage.

UK	European	Harmonised
240Vac	220Vac	230Vac +10% to -6%
415Vac	380Vac	400Vac +10% to -6%

Here's an interesting fact – despite the new "230V" label, the actual supply voltage hasn't really changed. It's a bit like renaming a street without moving any of the houses. And since there's no real incentive for electricity supply companies to adjust the supply voltage, our modern equipment has risen to the challenge. These devices are designed to handle a range of 230V +/-10%, which means they can cope with anything from 207V to 253V. Now, there's a twist. It is clear that any electrical equipment that's pushed to work at higher voltage is like an overworked athlete – it might run faster for a while, but it's likely to have its working life reduced by up to 46%. Now, focusing on the UK, you might notice that the voltage level is generally higher. This peculiarity is exactly why reducing the voltage can be quite a smart move. Imagine it like adjusting the sails on a boat to catch the wind just right, we can make the system more efficient and generate energy savings.

Where the voltage level is higher there is an opportunity to save energy thanks to the voltage optimization devices. During my electrical studies, I learned about Ohm's Law and its applications but often I only use a simplified version of it. For example, the simplified equation, $W = V \times A$, helps to calculate suitable cable sizes for different scenarios etc. However, I must admit that I misunderstood some parts of it. One common mistake is assuming that the wattage of a device always stays the same. This misconception is reinforced by the wattage information printed on equipment, like a 50W lamp. I used to think that if you lower the voltage, the current would automatically increase based on the simple formula I learned. Looking back, I now realize the importance of clarifying these misunderstandings and deepening my consideration of Ohm's Law. It's a reminder that even basic principles need ongoing examination and comprehension in the world of electrical engineering. In reality, the 50W rating of the lamp happens only when the voltage precisely matches 230V; in fact, with the average voltage in the UK being around 240V, the lamp would operate at over 55W, generating more heat and significantly shortening its lifespan. It's important to note that the physical resistance of the wire element in the lamp remains constant, while the wattage is actually dependent on the voltage. To accurately calculate the power, we must employ the formula represented in the Ohm's Law Circle: $P = V^2/R$.

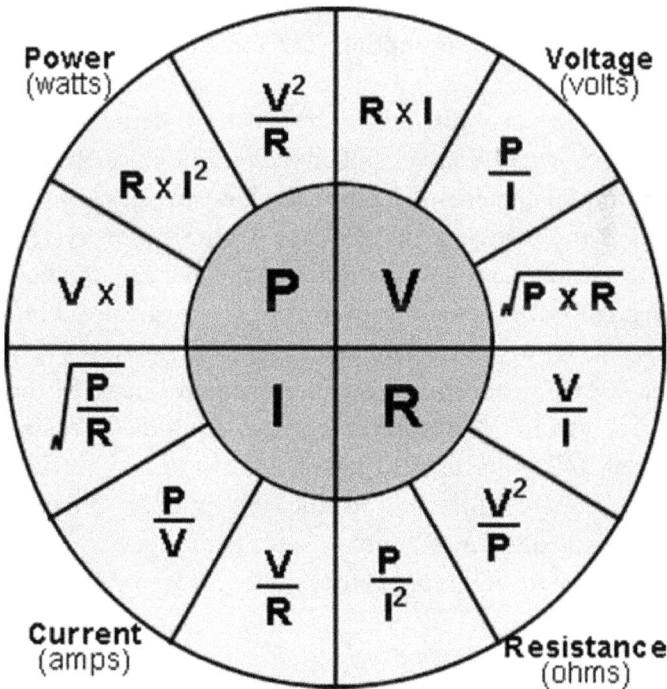

With equipment that is 100% voltage dependent, such as a typical 3kw electrical immersion heater element, every 1% reduction or increase in voltage results in a corresponding decrease or increase of 2% in power (watts). This relationship between voltage and power is worth examining more closely to better understand its implications for these types of devices. For instance, we know that the element, at 220V, will have 3000W of power (as per the element specs), which means 13.6A drawn. Let's see what happens if the voltage supply changes to UK at 242V:

Step 1:
Calculate the constant resistance: $220^2/3000 = 16.13\ \Omega$

Step2:
Calculate the Current drawn: $242/16.13 = 15A$

Step 3:
Calculate power consumption: $242^2/16.13 = 3{,}630W$

As we can see, the above calculation demonstrates how, depending on what voltage is applied to the same piece of equipment, the consumption, the costs and even the lifespan can vary.

This is the principle that Voltage Optimization systems use, applying Ohm's Law and optimizing the voltage of the power supply, significant savings of up to 10% can be achieved instantly. This translates to immediate reductions in electricity bills and a reduced carbon footprint. Furthermore, in line with the UK's wiring regulations (BS7671), electrical equipment designed to operate at 242 volts may experience a shortened working life; therefore, voltage optimization not only provides financial and environmental benefits but also ensures the longevity of electrical equipment according to regulatory guidelines.

Basic concepts of voltage in an electrical system

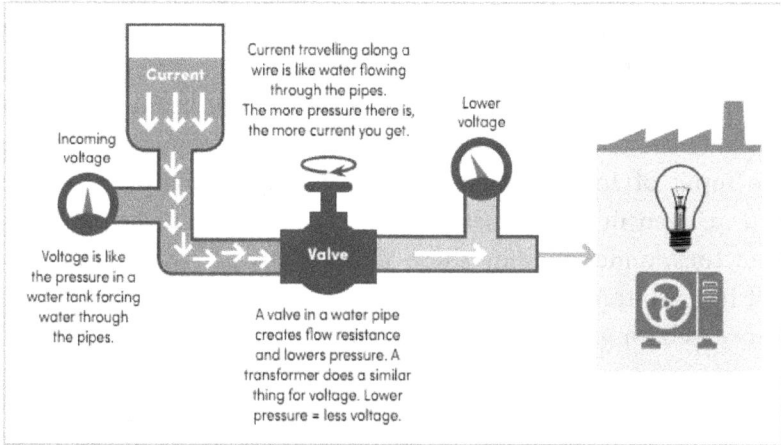

Current travelling along a wire is like water flowing through the pipes. The more pressure there is, the more current you get.

Incoming voltage

Lower voltage

Current

Valve

Voltage is like the pressure in a water tank forcing water through the pipes.

A valve in a water pipe creates flow resistance and lowers pressure. A transformer does a similar thing for voltage. Lower pressure = less voltage.

Image source: leadingedge-energy

Voltage optimizers are installed in series with your electricity supply, positioned between the distribution transformer and the main low-voltage distribution board. As the electricity flows through the voltage optimizer, it efficiently reduces the supply voltage to the desired level. Any excess voltage beyond the required amount is effectively returned back to the grid, ensuring it is not utilized on-site. By simply rejecting and returning the surplus voltage to the grid, your electrical equipment receives the advantage of an optimized power supply. This process allows for improved energy efficiency and ensures that your electrical devices operate with the ideal voltage levels for optimal performance.

In our complex buildings, one of the big consumers is the motor. These machines are important as they give power to many things inside a building. For example, when we use lifts to go up and down floors, motors are doing the hard work. They move the lift smoothly and safely. Then, there's the system that keeps our rooms warm or cool, the fans in the HVAC system. Motors help run this. Also, when we open a tap to get water or flush a toilet, motors are at work again. They help push the water up, even in very tall buildings, so we can use it whenever we want. In short, motors are like the hidden helpers in our buildings. As we navigate the complex world of motors, we simply cannot bypass the topic of International Efficiency (IE) standards. Whether you've been in the industry for decades or you're just dipping your toes, understanding these standards is paramount. They're not just arbitrary classifications; they mark pivotal indicators in our search for greater motor efficiency.

IE1 (Standard Efficiency)

Our journey begins with IE1, representing standard efficiency. For a long while, these motors were the standard in our industry. However, with growing energy concerns and a shift towards more environmentally conscious practices, it became clear that 'standard' needed an upgrade.

IE2 (High Efficiency)

Enter IE2, denoting high efficiency. Motors conforming to this standard offered a noticeable improvement over their IE1 counterparts. In practical terms, this meant reduced energy consumption, leading to cost savings in the long run. For businesses, this was not just a nod to eco-friendliness, but also a sensible economic choice.

IE3 (Premium Efficiency)

Taking a step further, we have IE3, signifying premium efficiency. These motors are designed with advanced engineering techniques to reduce energy losses even more. If you've ever wondered how industry leaders manage to cut down their energy bills significantly, IE3 motors play a significant role in that achievement.

IE4 (Super Premium Efficiency)

Lastly, the pinnacle of our current efficiency journey – IE4, or super premium efficiency. These motors represent the zenith of our current engineering capabilities. Incorporating cutting-edge materials and design techniques, IE4 motors are the epitome of performance, efficiency, and environmental responsibility.

The invention of the induction motor goes back more than 100 years. While several people contributed to its development, Nikola Tesla is often credited as its invention. He was the first to file for a patent in the United States in 1887. Imagine the motor as an elaborate dance with two main performers: the stator and the rotor. Understanding their roles and interactions is key to grasping the motor's operation.

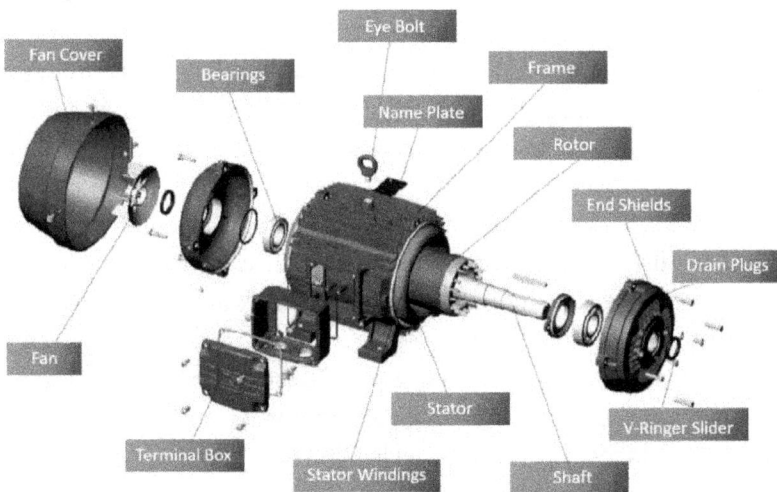

Image source: Flaktgroup

The Stator

True to its name, is static and does not move. Picture a cylindrical chamber. Within this chamber is a carefully crafted collection of electromagnets. To achieve this configuration, thin steel or iron layers are meticulously stacked together in a slotted arrangement to form the cylindrical shape (windings). But how do these electromagnets come into play? The magic happens when copper wire, wound in alternating directions, threads its way through the interior of this cylinder. The result is a creation of magnetic poles. Once alternating current, the lifeblood of our AC motor, courses through these wire coils, a fascinating transformation occurs. These coils form pairs of poles that alternate between north and south. And due to the alternating nature of the current, the polarity of these poles is in a perpetual dance, switching back and forth with each half cycle. The culmination is a magnetic field that doesn't just sit there – it rotates with a consistent strength, setting the stage for the rotor's entrance.

The Rotor

The second lead performer in our motor's ballet. Much like the stator, the rotor is also composed of a series of electromagnets, but with a twist – these are arranged in a manner resembling an axle, sitting snugly within the stator's embrace. As the stator works its magic, producing its rotating magnetic field, the rotor isn't just a passive observer. The magnetic fields created within the rotor are irresistibly drawn towards the stator's field. The rotor, in essence, tries to "catch" the stator's field, attempting to follow its rotation as it changes with every half cycle of the alternating current. This intricate interaction between the stator and the rotor gives the AC induction motor its name. The stator, by producing its rotating magnetic field, "induces" a magnetic field in the rotor. And here's a titbit to remember: unlike some other motor types, the rotor of an induction motor doesn't house any permanent magnets.

Continuing our exploration into motors, I want to talk about and introduce you to a personal favourite of mine: the EC fan technology. EC stands for Electrically Commutated, it harmoniously marries the best elements of AC and DC voltage worlds. One of the standout features of EC fans is their keen intelligence when it comes to speed control, paving the way for substantial energy savings. This isn't just a mere statement; there's science behind this efficiency. EC electric motors are innovative, employing a permanent magnet in their secondary field (primary field is the Stator field and secondary field is the rotor one). When the rotor springs into action, energised, it forms a repelling magnet, facilitating the motor's rotation within the magnetic field. The result? A whopping 30% energy savings over conventional motors, mainly because the secondary magnetic field sidesteps the need to draw energy to form its own magnetic field. Now, let's discuss a practical application. EC motors are designed for compactness, and with their inbuilt controller solution, they achieve optimal cooling effects, courtesy of the surrounding airstream, below some features:

Future-Proofed: They're all set for ERP 2018 (energy-related products) requirements, ensuring a sustained relevance.

Conserving Energy: On average, these motors save about 30% electricity in comparison to traditional fans, reflecting their eco-friendly DNA.

Integrated inverter drive, significant noise reductions are achieved.

Built-in Protection: Say goodbye to additional motor protection such as harmonics filters. EC fans come equipped with their own protective mechanisms against overcurrent and temperature spikes.

BMS Compatibility: They integrate effortlessly with Building Management Systems.

Durability and Maintenance: They're not only easy to maintain but also promise a long service life.

Space-Efficient Design: The external rotor motor concept ensures they're compact and space-saving.

Integrated Diagnostics: The inbuilt status LED eliminates the need for separate displays or tools for failure diagnostics.

Another intriguing new technology is the "newborn" Smart Motor. These motors represent a significant jump in both design and functionality, pushing the boundaries of what we've come to expect from traditional motors. Smart motors are more than just machines; they encapsulate the integration of traditional motor mechanisms with modern digital systems. These motors have the inherent ability to self-optimise, adapt to changing conditions, and synchronize seamlessly with other digital infrastructures. The result is unparalleled energy efficiency, sophisticated automation, and enhanced system control. While smart motors are based on established motor designs, their true distinction lies in the infusion of contemporary digital control techniques and software. This allows for meticulous and precise control, thereby mitigating many inefficiencies associated with conventional motors.

Core Design: Unlike traditional designs, these motors utilise a unique setup where the rotor doesn't contain conventional copper windings. Instead, stator windings produce a magnetic field, with the rotor being driven by alterations in this magnetic field. Some smart motors are based on the switched reluctance principle. In these motors, the rotor is made of laminated iron without any windings. The stator, which surrounds the rotor, contains wound coils. When a coil on the stator is energised, it magnetically attracts the nearest piece of the rotor, causing the rotor to turn. By sequentially energising the stator coils, the rotor can be made to turn continuously. Just recently, the technology necessary to control this process has been developed.

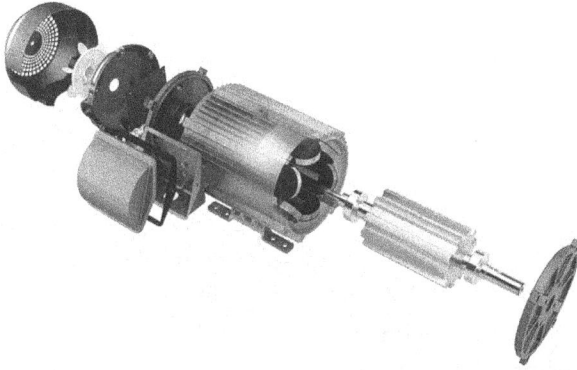

Image source: FutureMotors

Digital Precision: Advanced sensors and software ensure that these motors operate at peak efficiency at all times. The system can dynamically adapt, curtailing any superfluous energy consumption and ensuring optimal operation regardless of external influences or shifts in load.

The future of motors is not limited to hardware alone. Integration and connectivity are paramount in today's digitised world. Embedded sensors and connectivity capabilities enable continuous performance monitoring of these motors, paving the way for predictive maintenance and timely interventions.

Cloud synchronisation: By connecting to cloud platforms, the data analytics opportunities arise, granting businesses invaluable insights to refine operations and further decrease energy expenditures.

Advanced management Systems: These motors can be effortlessly integrated into building or industrial control systems, ensuring cohesive operation with other smart devices and infrastructures.

A co-conscious approach: The advent of smart motors signifies a stride towards a more sustainable future. With their heightened efficiency and reduced energy wastage, they contribute significantly to the reduction of greenhouse gas emissions. Some reports suggest that, compared to traditional counterparts, smart motors can save up to 40% of energy. We stand at the threshold of a transformative period where the confluence of age-old engineering principles and cutting-edge digital technology promises to redefine our understanding of machinery. Smart motors are showing the way towards an era where efficiency and sustainability are intrinsically intertwined.

As we explore how the world of engineering is linked with energy efficiency, mention of the Variable Speed Drives (VSDs) is a must. These clever devices help control the speed of motors, fans, and pumps, making sure they match the specific needs of an application. The best part? They can lead to significant energy savings! So, how do they work? In more technical language, VSDs transform the fixed frequency and voltage of the incoming electrical supply into a variable frequency and variable voltage output. This results in a change in the motor's speed and torque. But don't worry, the VSD has got it all under control. It can manage the motor's speed from zero RPM all the way up to about 100-120% of its full rated speed. And here's the kicker: you can achieve up to 150% rated torque at reduced speeds. Imagine this: by using a Variable Speed Drive to decrease a fan or pump motor's speed from 100% to 80%, you can save up to 50% of the energy consumed. Impressive, right? In fact, every fan, pump, or HVAC motor could benefit from a VSD, as it significantly reduces electricity usage by matching the actual requirements instead of always running at full speed. Even the Carbon Trust[*] acknowledges the advantages of Variable Speed Drives in cutting down electricity usage, costs, and carbon emissions.

---

[*] *The Carbon Trust in the UK is an independent, expert organization focused on supporting the move towards a sustainable, low-carbon economy. Established in 2001, the Carbon Trust works with businesses, governments, and institutions around the world, offering advice, expertise, and support to reduce carbon emissions, increase resource efficiency, and drive the adoption of low-carbon technologies. Their services include providing guidance on carbon reduction strategies, energy efficiency improvements, and renewable energy solutions. The Carbon Trust also offers certifications for organizations that meet specific carbon reduction and sustainability standards. By helping organizations identify and implement energy-saving measures, the Carbon Trust plays a vital role in mitigating climate change and supporting the transition to a more sustainable global economy.*

This reduction in speed directly translates to decreased power and energy usage. Variable Speed Drives offer substantial electricity savings and can be developed in a range of applications. A thorough site survey can help identify the best application and anticipated energy savings. It might sound simple – and that's because it is! VSDs are both simple and effective solutions for energy savings.

How do variable speed drives regulate power to match the motor's speed to what is required by the process?

1. Convert incoming AC to DC

Incoming 3-phase AC power is fed into a rectifier that converts it to DC power.

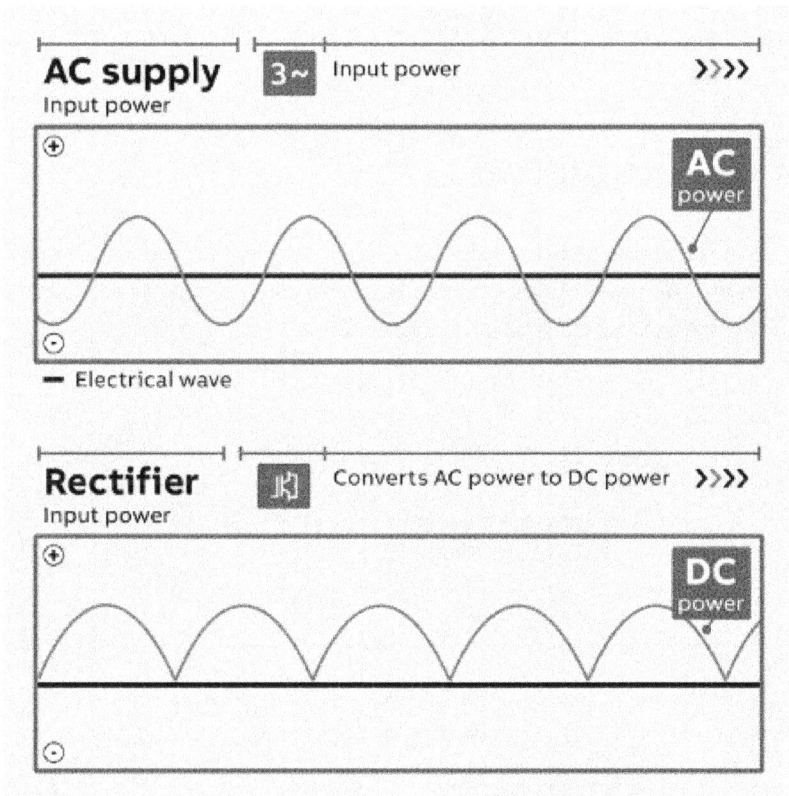

AC supply 3~ Input power >>>>
Input power

AC power

— Electrical wave

Rectifier Converts AC power to DC power >>>>
Input power

DC power

2. Smooth the DC wave

DC power is fed into capacitors, smoothing the wave, and producing a clean DC supply.

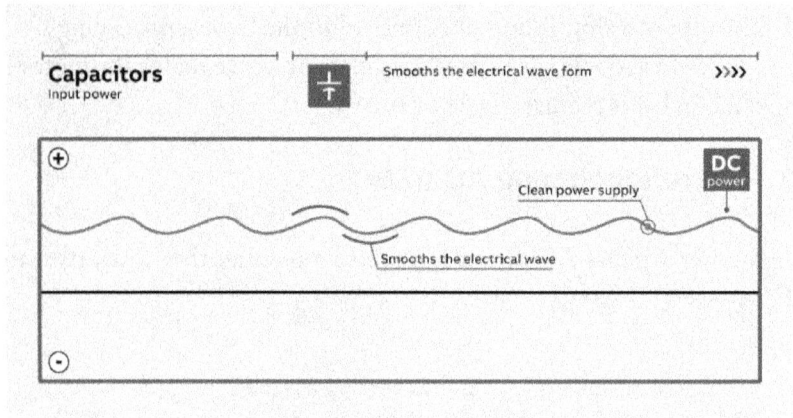

3. Convert DC to variable AC

The variable speed drive calculates the motor's required voltage and current. DC power is then fed into an inverter producing AC power at the precise voltage and current needed.

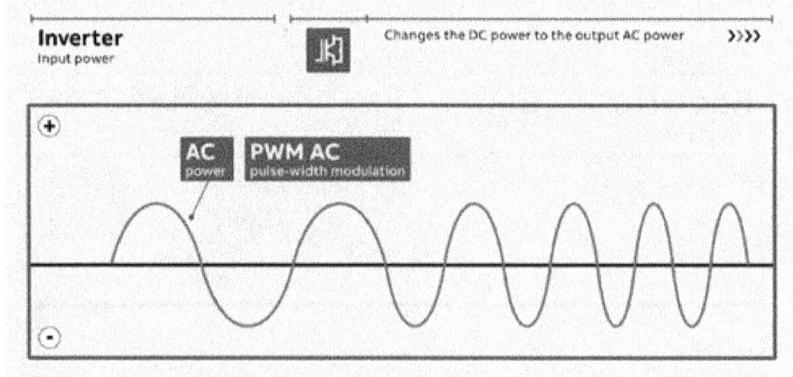

4. Calculate and repeat

The variable speed drive continuously calculates and adjusts the frequency and voltage providing only the power (speed and torque) the motor needs. This is how you can save large amounts of energy.

To give you a better idea the below graph shows the impact of a retro-fit VSD installation to a kitchen extract fan with a motor power of 4Kw which is in operation for 8,760 hours per year (for 24-hour, 365-day operation).

TOTAL ENERGY CONSUMPTION

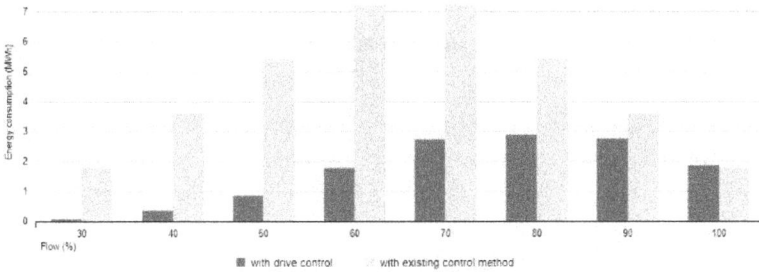

with drive control   with existing control method

RESULTS

22.8 MWh
Annual energy savings

36.2 MWh
Annual energy consumption with existing control method

13.4 MWh
Annual energy consumption with drive control

63.1 %
Annual energy savings percentage

3,421 €
Annual energy savings

6.8 t/year
CO_2 reduction

Here's why investing in Variable Speed Drives (VSDs) can boost the energy efficiency of your property, including pump, fan, conveyor, and compressor systems:

1. Enhanced operational efficiency
2. Lower electricity bills

3. Reduced capital expenditure, an immediate decrease in electrical consumption ensures a quick financial return on your VSD investment. In pump and fan applications, this payback can happen even within months of installation.
4. Savings on maintenance and spare part costs, using drives minimizes the strain on mechanical equipment during start-up and operation, ensuring a longer lifespan for your equipment and ultimately saving on maintenance and spare part costs.

Synchronous Reluctance Motors (SRMs) are another exciting development in the field of motor engineering. They use complex geometry to maintain low internal resistance in all motor shaft positions, resulting in improved fluidity and motor output. Unlike AC induction motors, SRMs can deliver superior low-speed torque and efficiency while being more compact in size. Magnetic bearings are another game-changing technology in the world of motors. Using a system of active electromagnets, they can support spinning shafts without any physical contact, resulting in reduced friction, heat, and vibration, and increased reliability. As the demand for high-efficiency motors increases, researchers are exploring rare-earth replacements in combination with alternative motor configurations that retain the efficiency. VSD software is also rapidly evolving with self-learning and modelling capabilities that can minimize the need for sensors as inputs into the control system.

Newest VSDs offer internet connectivity, allowing for rapid analysis of system performance and predictive scheduling of equipment maintenance and repairs. With a single point of connection for a multitude of sensors and data points, operators can receive notifications on their smartphones before a failure occurs. I believe that all these exciting innovations in motor engineering are just the beginning of what is yet to come. As the sector continues to evolve, engineers can look forward to new and exciting advancements that will shape the future of our world.

Demand Controlled Ventilation

A system which has a strong connection with motors is the demand-controlled ventilation, specifically looking at Demand Controlled Ventilation (DCV) systems. Picture this: Instead of a one-size-fits-all approach like the Constant Air Volume (CAV) systems, where airflow is constant, DCV systems are like the intuitive geniuses of ventilation. They adjust the airflow to match exactly what's needed at any given moment. Think about it this way. If you're in a room that's empty, does it really need the same amount of air as when it's full of people? Of course not! That's where these smart systems come into play. In a room with no one around, they'll scale back the air supply. But if another room is bustling with activity, they'll crank up the volume of air to that area. It's all about being efficient and meeting the needs of the space. Let's explore how these systems make real-time decisions to provide just the right amount of ventilation, ensuring comfort, efficiency, and energy savings.

When it comes to crafting the perfect indoor climate, there's a symphony of factors at play, each as crucial as the next. Multiple elements come together to create a workplace that's not just liveable, but comfortable and healthy. Let's take a closer look at these key players:

Lights: Think of lights as the stage spotlight. They set the mood, influence our sleep patterns, and even impact our productivity. Whether it's the warm glow of a lamp or the bright light of the midday sun streaming in through a window, lighting plays a pivotal role in how we perceive and enjoy a space.

Humidity: This is all about the air's moisture content. Too dry, and you might find yourself reaching for the lip balm more often. Too humid, and it's like living in a steam room. Getting the humidity just right is crucial for comfort, health, and even for preserving the integrity of the building materials and furnishings.

Thermal Comfort: Not too hot, not too cold. It's a delicate balance influenced by factors like air temperature, radiant temperature, air speed, and humidity. It's about creating a climate where you can sit back, relax, and say, "Ah, this feels just right."

Indoor Air Quality: Imagine a breath of fresh air – that's what good indoor air quality feels like. It's about having clean, refreshing air to breathe, free from pollutants, allergens, and odours. Proper ventilation is key here, ensuring a steady supply of fresh air while whisking away the stale air.

Air Flow: Think of air flow as the gentle breeze on a calm day. It's about how air moves around a space, contributing to thermal comfort and air quality. It's not just about the quantity of air, but also its distribution and speed.

Acoustic Environment: Last but not least, the acoustic environment is the soundtrack of your space. It's about controlling noise levels and ensuring sound quality. Whether it's the hum of an air conditioner, the chatter from the next room, or the echo of footsteps, the acoustic environment can deeply impact comfort and concentration.

Now that we've got a solid understanding of the key elements influencing climate comfort in a building, it's time to tackle the next big challenge: efficiency. It's like being a smart chef in a kitchen, knowing exactly what ingredients to use to create a delightful dish without any wastage. When we talk about efficiency in the context of indoor climate, we're essentially looking at how to satisfy the needs and preferences of the people using the space. It's a bit like being a detective, figuring out what everyone in the room needs and then crafting a plan to meet those requirements in the most effective way. How do we measure comfort in a building? What's the relationship between the level of comfort we're aiming for and the investments we're willing to make for the equipment? Finding the right balance is the key. Imagine you're on a tightrope, and on one side is the max comfort, and on the other side are the costs and energy considerations. Your goal is to walk this tightrope with balance and skills. Now, here's where it gets interesting. Technical solutions for maximizing climate comfort come in many shapes and sizes. The secret? Knowing which tool to use and when.

The concept behind adjusting airflow in buildings is straightforward, yet it's a game-changer in how we approach ventilation. Imagine you have a water tap. You wouldn't keep it running at full when you only need a trickle, right? The same logic applies to ventilating rooms. With traditional constant air volume systems, the airflow is like that fully open tap – it's constant. These systems often operate at a higher volume, primarily to ensure that there's enough air supply when a room is packed with people. But here's the catch: this 'one-size-fits-all' approach doesn't change even when the room is completely empty. It's like keeping the tap running full force, even when you're not using the water. This is where the concept of demand-controlled ventilation comes into play, transforming how we think about the use of air in our buildings. It's all about customization and efficiency. Variable Air Volume (VAV) and Demand Controlled Ventilation (DCV) systems are the protagonists. They're like smart taps that adjust the flow based on how many people are in the room. So, what's the big deal? These systems intelligently adjust the air supply to match the occupancy of the room. Empty room? Less air. Full house? Increase the flow. It's a simple yet effective strategy that ensures you're not wasting energy on unoccupied spaces.

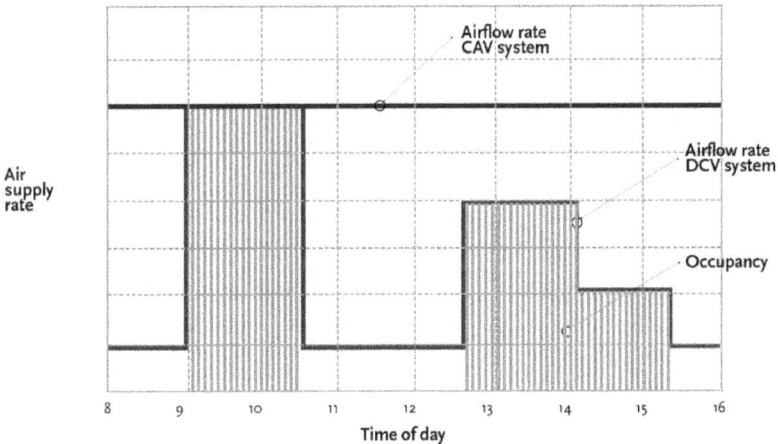

Let's break down how DCV does its thing, shall we? It's a bit like a clever thermostat, but for your building's air. Here's what it's keeping an eye on:

Temperature: It's all about maintaining that 'just right' feeling in a room. Not too hot that you're doffing layers, and not too cold that you're reaching for a jumper. The system adjusts to keep the temperature spot on.

Indoor Air Quality: This is becoming a crucial topic and an essential requirement for any ESG's organization (Environmental, Social and Governance) standards. We want the air indoors to be as fresh as a daisy, comfortable to breathe, and healthy too. If the DCV gets a scent of anything less than pleasant, it steps up to sort it out.

Humidity: Now, this one's important for the building itself. Too much moisture in the air and you might as well send out invites for mould to move in. The DCV keeps humidity levels in check to avoid this damp squib.

Occupancy: As mentioned, there's no sense in pumping loads of air into an empty room. This efficiency isn't just good for your energy bills; it's also a golden ticket to a top-notch classification in various green building certification programmes like BREEAM, WELL, and LEED. These are the badges of honour in the building world, they confirm your reputation in sustainable practices, proving that your actions speak louder than words.

How it works?

DCV systems typically comprise sensors that detect occupancy and temperature, coupled with a control unit that commands the ventilation machinery accordingly. Here's a look at the ways in which a DCV system might measure occupancy and temperature within a building:

CO_2 Monitoring: DCV systems may include CO_2 sensors distributed throughout the premises. These sensors monitor the carbon dioxide levels, which naturally rise as more people populate a space due to respiration. The system uses these readings to adjust the ventilation rate to ensure adequate air quality.

Occupant Counting: Some systems use direct methods to count how many people are present, employing mechanisms such as ticketing data, entry card swipes, or even visual recognition technologies.

Occupancy Detection: Other systems use specialised sensors or a combination of different sensors analysis to determine if an area is occupied.

Temperature Measurement: Certain DCV systems also include temperature sensors to help modulate the ventilation based on the thermal readings of a space, ensuring the indoor climate remains within a comfortable range.

Fitting out new buildings or full refurbishment projects with Demand Control Ventilation systems is a breeze, but what about retrofit for older buildings? Can they keep up with this modern tech? The good news is, they certainly can. Today's DCV systems are quite the chameleons, able to blend into older buildings. So, regardless of a building's age, DCV systems can give it a new lease of life, ensuring that it breathes right in today's world, balancing the comfort of occupants with energy efficiency. I always find it fascinating to bring the past into the present and about that, I have been involved in a DCV system which is designed for kitchen ventilation. Demand Controlled Kitchen Ventilation systems, or DCKV, come in a variety of styles. In this case, although the principle and the goal remain the same, is not the occupancy the main aspect to consider. The classification of these systems is based on their method of detecting heat and/or cooking activity. There are different types of DCKV systems which includes multiple sensors to collect the necessary data to manage the extraction and supply efficiently and automatically. These types of systems use a resistance temperature detector (RTD), often in combination with opacity sensors (This is a reflective beam in the canopy. Steam and or smoke can be generated before the temperature set point is reached. In this way, if the smoke and/or steam block the "beam", it will automatically ramp the exhaust system to design airflow). The RTD is usually installed in the exhaust collar of the hood. It's like a sentinel, keeping a watchful eye on the temperature. Some systems go a step further, placing multiple RTDs within the canopy of the hood. This way, they ensure that heat is detected evenly across the entire length of the hood, leaving no hot spot unnoticed. Once the temperature in the hood hits this pre-set threshold, the RTD sends a signal to the exhaust fan's VSD. This is where the exhaust fan adjusts its speed to match the cooking activity. The interesting part is how the system decides how much air to exhaust. This decision is based on the system's settings for two distinct phases: small and active cooking. During small cooking, when the appliances are on but not actively used for cooking, the system maintains a lower air exhaust rate. When the actual cooking begins, the system ramps up the exhaust rate, responding to the increased heat and cooking activity. Simultaneously a signal that is proportional to the exhaust frequency is sent to the corresponding supply air unit to keep the system balanced. Obviously the designers and installers should take extra caution in determining this output signal because external factors should be considered such as gas safety interlock systems. More sophisticated systems commissioning a varied range of sensors. These include infrared sensors and thermal imaging sensors which are aimed directly at the cooking appliance surfaces. The infrared sensors are not just looking at individual readings in isolation; they're comparing data to get a full picture of what's happening in the kitchen, for instance, the scenario of dropping frozen fries into hot oil. As the overall temperature might remain relatively steady, the infrared sensor will notice a sudden drop in temperature at the surface of the cooking appliance. This isn't just a random temperature fluctuation; it's a clear sign of cooking activity. The system is designed to recognize such indications and responds instantly.

The thermal imaging sensors are an enhancement in the field of DCKV systems, offering a significant jump in both sensitivity and potential for energy savings. This advanced sensor utilizes a grid-style thermopile, which boosts both resolution and thermal sensitivity, allowing for a greater understanding of the kitchen environment. One of the standout features of this sensor is its wide 110° viewing angle. This means that a single sensor can effectively monitor a large area, suitable for hoods up to four metres in length. The thermal imaging sensor provides a detailed thermal portrait of the cooking area. It delivers a remarkable number of individual temperature readings, enabling the DCKV system to ascertain with precision what's happening with each appliance under the hood. This granularity of data allows for air volumes to be adjusted more accurately based on actual cooking activity, rather than just relying on average temperature readings. The response time, it's almost instantaneous and the system reacts to changes in cooking conditions in real-time. This means that the ventilation system can adjust swiftly and precisely to the immediate needs of the kitchen, ensuring optimal air quality while maximizing energy efficiency.

Effectiveness

Imagine a heating or cooling system that runs perfectly, using every bit of energy it's given to do something useful without wasting any. That's what we mean when we talk about 100% energy efficiency. It's like a really good athlete who uses all their energy to win the race and doesn't waste any just running in circles.

Just like in real life, it's pretty hard to find a system that's this perfect. Things like friction, heat escaping, or just general system inefficiencies often get in the way, making it a tough goal to reach. You might find it surprising, but some systems can appear to beat themselves, showcasing an efficiency that exceeds 100%! This is due to our methods for defining and measuring efficiency. Let's take a look at heat pumps as an example. Heat pumps are exceptional because they operate by transferring heat, not creating it. They're able to pull more thermal energy from the surroundings, like air or soil, and deliver it to a building. This means they actually give back more than they take, making their efficiency higher than 100%! The term 'Coefficient of Performance' (COP) describes this impressive characteristic. In terms of fuel-burning heating systems, some high-efficiency models can reach efficiencies in the 90-98% range, but none can exceed 100%. Condensing boilers can achieve efficiency levels of around 90-98%, and can sometimes be described as having an efficiency greater than 100% when measured under specific conditions. This is due to the way they utilize the latent heat of vaporization which would normally be lost from the flue gases in a conventional boiler.

But... Hey! what's the COP?

It's an important tool used in thermodynamics to measure and quantify the efficiency of energy conversion and transfer system. The COP is like a report card for your system performance, showing how energy efficient it is. Think of it as a scale of heating or cooling provided by your air-conditioning unit compared to the electricity it needs to get the job done. Let's make this more concrete with an example.

If your air system can generate 5kW of heat using just 1kW of electricity, its COP is five. But if it only produces 2.6kW of heat with 1kW of electricity, then its COP is 2.6.

These measurements, which don't have any units, are used all around the world to help you understand how efficient your equipment is. To put it simply, the higher the COP the better your equipment is at saving energy. An HVAC system with a COP of at least three is like a star engineer, it's giving back more energy than it needs to run. If you aim to control your energy usage effectively, opting for an HVAC system with a higher COP is a wise decision. This choice will help you maintain lower bills while maximizing energy efficiency.

Now let me walk you through the Combined Heat and Power (CHP), a pretty remarkable system that's quite the multitasker, producing both electricity and thermal energy with stunning efficiency. Imagine a symphony of varied technologies and fuels, uniting to generate power on-site. This synergy mitigates energy losses and harnesses the otherwise escaping heat, repurposing it for practical uses such as process heating, steam creation, and producing hot water. CHP's ability extends far beyond singular building applications. It reveals its versatility in environments as diverse as district energy systems, microgrids, or even as a utility resource, supplying power and thermal energy to a multitude of users. In addition, CHP is a symbol of reliability, ensuring uninterrupted power round-the-clock, even within a power outage. It coexists harmoniously with other distributed energy technologies, including solar photovoltaics (PV) and energy storage systems. Should we compare CHP to conventional power and heat generation systems, the difference in energy efficiency is clear. Traditional systems waste nearly two-thirds of energy, dissipating it as heat during the processes of generation, transmission, and distribution. In contrast, CHP seizes this potentially wasted heat, sidesteps distribution losses, and consequently elevates its efficiency above 80 percent.

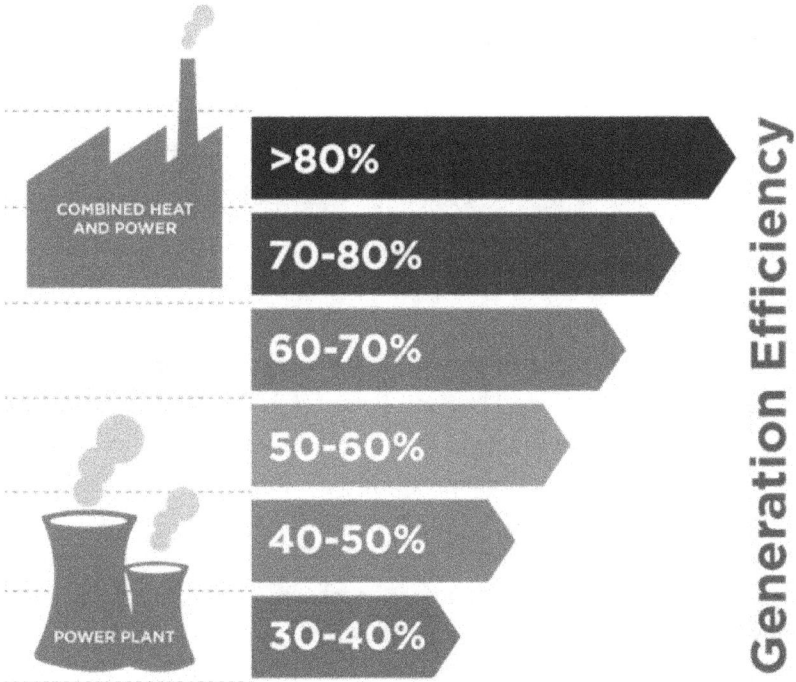

The CHP can work in conjunction with an absorption chiller providing electricity, heat and cooling. An absorption chiller is a type of cooling system that operates using heat, rather than electricity (compressor), which is commonly used in traditional refrigeration systems. It works by applying heat (hot water or steam) to a refrigerant, causing it to evaporate and absorb heat from its surroundings, thus creating a cooling effect. The evaporated refrigerant is then reabsorbed, and the cycle begins again. This type of chiller is particularly useful in situations where waste heat is available.

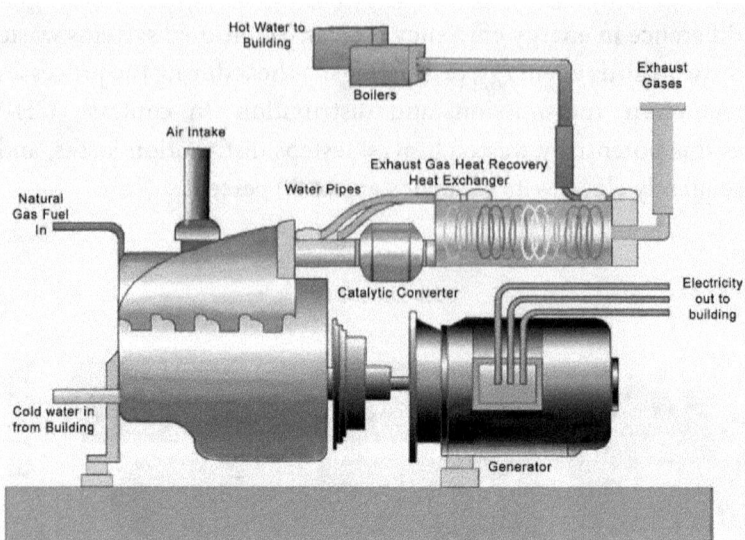

The picture above is a simplified visualization of the primary elements of a Combined Heat and Power (CHP) unit. Let's begin on the left, where you'll find the 'prime mover'. Often this is a diesel engine adapted to run on gas, usually natural gas. The prime mover sets the generator in motion, producing the electricity needed for the site. The wires from the CHP unit run parallel to the incoming power lines from the utility provider; therefore, according to the principles of physics, locally produced energy is consumed first, with the utility provider's power filling in any gaps to meet the site's energy demands. Thermal energy is reclaimed from the engine jacket and the exhaust gases via an exhaust gas heat exchanger. This heat exchanger is usually a shell and tube design where water circulates around tubes carrying the exhaust gases. Despite my admiration for CHP's formidable efficiency, it isn't my ideal solution. It continues to rely on gas for operation, and my personal vision aligns more closely with purer forms of energy generation. Some propose that CHP could function with hydrogen, but that concept, while promising, remains unfeasible at present.

Another common CHP system configuration is the CHP Steam boiler (with steam turbine). In this case the principle is the same but instead of an engine the process begins by producing steam in a boiler. The steam is then used to turn a turbine to run a generator to produce electricity. The steam leaving the turbine can be used to produce useful thermal energy. These systems can use a variety of fuels, such as natural gas, oil, biomass, and coal.

Below configuration:

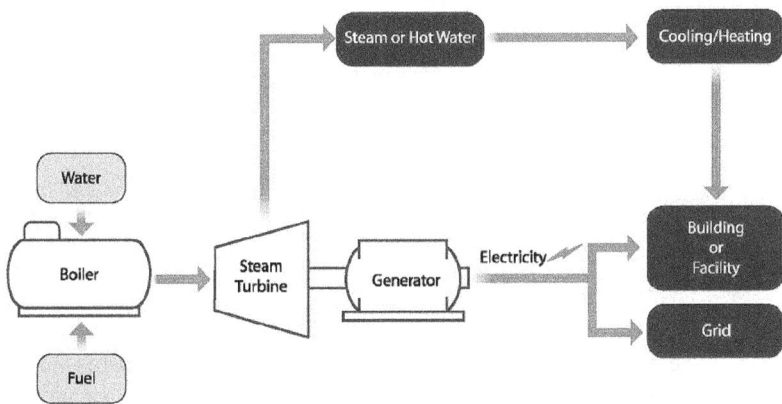

There are different ways to integrate a CHP unit, as already mentioned using a parallel interface or systems which utilise a thermal store or series connection. The below series connection setup shows the CHP unit taking a portion of the cooler returned water, the CHP heats it, and then mixes the heated water with the remaining returned water before directing it to the existing set of boilers. One of the main advantages of this type of interface is that the CHP unit is always the lead boiler by a virtue of it seeing the coldest water first. This configuration also means that the need for a BMS control system is virtually non-existent, making it simple and ideal for older buildings without a large BMS system.

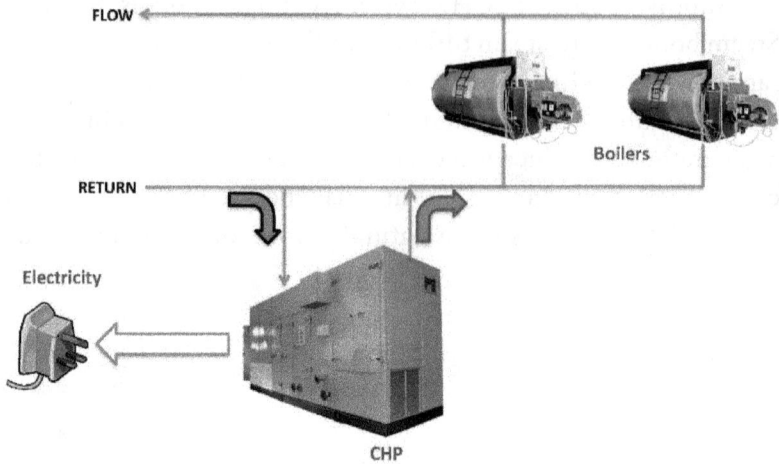

FLOW

Boilers

RETURN

Electricity

CHP

The Economics of the CHP

While the attraction of Combined Heat and Power systems lies in their potential for efficiency and environmental benefits, it's vital to understand that these systems do not always equate to cost savings. This chapter aims to demystify the financial considerations associated with CHP systems, emphasizing two major components: utility prices, including the Climate Change Levy (CCL), and the CHP operational data.

- **Utility costs and the Climate Change Levy:**

The first significant factor in the financial equation of a CHP system is the utility price. This includes the costs of fuels, typically natural gas, and the electricity prices that can be offset by generating power on-site. These prices can fluctuate significantly due to market dynamics, and considering the economic viability of a CHP system is essential.

1. Electricity rate (p/kwh)?

2. Gas rate (p/Kwh)?
3. CCL for both gas and electricity (p/Kwh)? (to calculate the costs savings in case of exemption)
4. Is your unit CHPQA certified? (If yes CCL exemption)

An important part of the utility price equation in many regions is the Climate Change Levy (CCL). This is an environmental tax on energy delivered to non-domestic users in the United Kingdom, designed to encourage energy efficiency and reduce carbon emissions. The Combined Heat and Power Quality Assurance Program, CHPQA, was introduced at the same time as the CCL (in 2001) to incentivise CHP installation by offering additional benefits such as CCL exemption and preferential business rates. To obtain these benefits your CHP unit must be registered and certified.

- **CHP Operational Data**

The other crucial component in understanding the financial implications of a CHP system is operational data. This includes details like the system's capacity, efficiency, expected lifespan, and maintenance costs, along with the heat and electricity demand of the building or site where the CHP system is installed. The ratio of heat to power demand can significantly impact the system's cost-effectiveness. In an ideal scenario, both the heat and electricity produced by the CHP system would be fully utilized, maximizing its efficiency and cost savings. However, if the demand for one form of energy is considerably lower than the other, it could lead to wasted energy and diminished returns. Furthermore, the operational and maintenance costs of the CHP system must be taken into account. These can include regular maintenance, replacement of parts, and unexpected repairs. Accurately accounting for these costs in the economic analysis can help avoid unexpected financial surprises down the line.

1. Maintenance contract costs (including call-outs forecast)
2. CHP hours run

3. CHP specifications
   - CHP Gas input (KWh)
   - CHP electricity output (Kwh)
   - CHP heat output (Kwh)
4. Heat efficiency of the "back-up" existing boilers (to calculate the heat that would have been supplied by the site boilers which would use gas to generate it, if the CHP was not available)

Once you have all the data above for a specific period you would like to check, then you can calculate the CHP savings. In conclusion, CHP, with its impressive energy efficiency, certainly deserves applause. But let's keep our eyes on the horizon, optimistic for the technological advancements that could soon present cleaner, more sustainable alternatives. CHP is indeed an important piece of the puzzle, but we should continue our search for the complete picture, the ultimate balance of sustainability and efficiency. Most properties are connected to the gas grid which means they can have a natural gas boiler that is supplied with fuel automatically. Properties which are not connected to the gas grid, e.g. those in rural areas, will need to use an alternative fuel for the boiler such as oil and electricity. Recently many buildings are making the choice to install a boiler or heating system which uses renewable fuel such as a biomass boiler or a heat pump in order to lower their carbon emissions.

Gas:

It is cheaper than electricity, making it a much more economical option when it comes to heating your property.

Oil:

If your property is not connected to a gas line, using an oil-fired boiler can be one of the most cost-effective alternatives. It is, however, more expensive than using gas, so if you have access to gas, it may be worth considering that option also. You can run out of oil, as it is not available on demand as for gas or electricity; instead it must be ordered and then stored in a tank. You will need to monitor the amount of fuel you have left and order it as needed.

Electric:

A typical electric boiler will work by heating the water that runs through it with a heating element and this hot water is then pumped to where it is needed. Despite being so efficient, electric boilers can be expensive to run due to the high cost of electricity and can only heat small amounts of water at a time. This means a larger property which uses a lot of hot water, or has multiple bathrooms being used at once may not find an electric boiler adequate.

Biomass:

Certain fuels used by biomass boilers are considered carbon neutral, making them much more environmentally friendly than traditional boilers. When burnt, the wood releases around the same amount of carbon it absorbed when it was growing. If you use locally sourced biomass this will reduce the carbon needed to transport it, making it greener; in fact, having independence from gas and electricity means you won't be affected by any price rises from suppliers and some biomass boilers are eligible to receive the government's Renewable Heat Incentive (RHI).

Hot water systems operating temperatures

Low Temperature Hot Water Heating System – LTW

- operates within a temperature of 120C. The maximum allowable working pressure for a LTW system is usually 2 bar.

- LTW systems are in general used for space heating in homes, residential buildings, offices, hotels, local distribution of district heating systems and similar.

Medium Temperature Hot-Water Heating System – MTW

- MTW systems operate at a temperature of 177C or less. The maximum allowable working pressure for a MTW system is usually 10.5 bar
- MTW systems are often used in large hot-water distribution systems like district heating, and in systems where process applications require higher temperatures than achievable by LTW systems

High Temperature Hot-Water Heating System – HTW

- HTW systems operate at temperatures exceeding 177C. The maximum operating pressure for a HTW system is usually less than 20.7 bar
- HTW systems are used like MTW systems in large district heat networks.

One of the great ways to save energy is to recover heat from hot water that would otherwise be wasted. Wait, what?! Yes, that was my reaction too. Let's start from the beginning. Wastewater Heat Recovery systems extract heat from the water used in showers, repurposing it to warm incoming mains water. These systems can be connected to both standalone showers and showers over baths, and commercial kitchens leveraging the energy that would otherwise be lost.

The process involves passing your used shower water through a heat exchanger.

How a Drain Water Heat Recovery System works

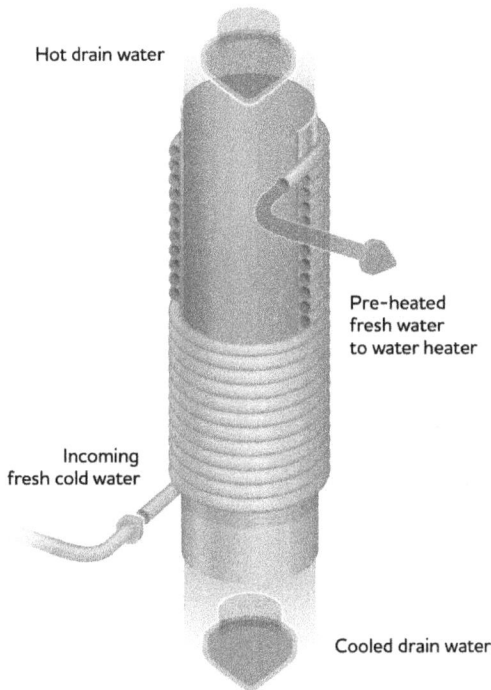

Hot drain water

Pre-heated
fresh water
to water heater

Incoming
fresh cold water

Cooled drain water

Image source:Mhydro

This captured heat is then utilized to pre-warm the cold feed of your thermostatic shower, making the system both efficient and eco-friendly. These devices operate at approximately 60% efficiency, meaning they can recover 60% of the energy that would typically be lost down the drain. This recovered heat is utilized to increase the temperature of the incoming cold-water feed from 10°C-12°C to 25°C-28°C. The pre-heated cold water can then be directed to the cold feed of the thermostatic shower and/or the hot water system (such as a cylinder or combi boiler). As a result, this reduces the demand for generated hot water for each shower, lowering the amount of energy required to heat water to the desired temperature. This not only reduces the strain on your boiler but also leads to significant energy savings. An efficiency rate of 60% is impressive, translating to a reduction in shower energy usage of 40%-50%. This offers substantial potential for savings on your hotel energy bills, while also promoting a more sustainable approach to managing your hot water consumption.

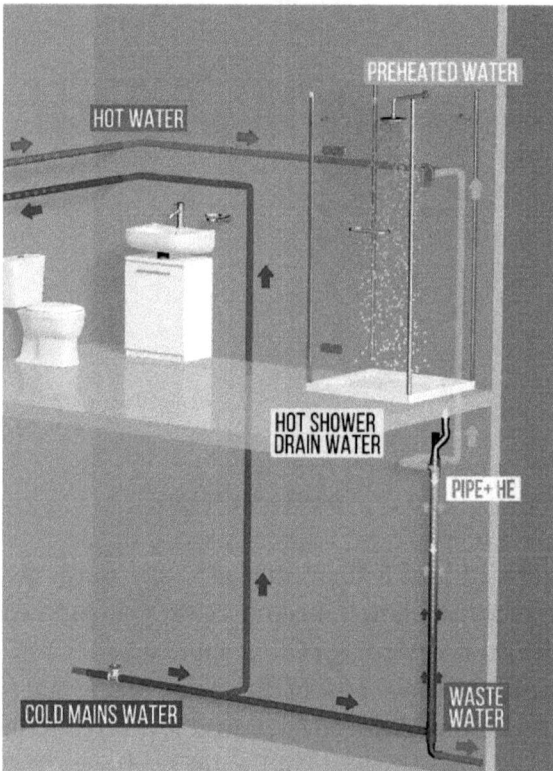

Image source:Recoup

The encouraging aspect of wastewater heat recovery systems is their simplicity. These devices lack electrical components, pumps, or controllers, which translates to minimal maintenance requirements. With an expected lifespan of more than 20 years, they are designed for longevity. Also, as a passive technology it is always working; in fact, there is no need to turn it on or off. However, their cost can be substantial, especially for retrofit projects. Fortunately, this scenario is starting to change.

Room's Energy Management System

Another great system I would like to talk about is the REMS system, which I think is ideal for hotels as it focuses on a simple but essential principle: mitigate energy consumption when a room or an area is unoccupied. Imagine walking into a room that immediately adapts to your needs, optimizing comfort while being mindful of energy use. At the heart of REMS is an intricate network of sensors and controls designed to detect occupancy. When a room is empty, the system automatically adjusts settings for HVAC, lighting, and appliances to minimize energy waste. It's not just about turning things off, but scaling down to optimal levels to ensure immediate comfort when the room is occupied again. The technical sophistication of this system comes from its ability to integrate seamlessly with a building's existing infrastructure. Motion detectors, door sensors, and even smart thermostats work in concert to provide real-time data on room status. This data is then processed through advanced software that adjusts energy output based on several factors:

Occupancy: REMS can determine if a room is empty or if it's being used, adjusting environmental controls accordingly.

Environmental Conditions: It considers external temperatures and natural light, tweaking indoor settings to maintain comfort without excessive energy use.

Operational Needs: Certain areas within a hotel may have specific energy needs based on their use or time of day, and it can adapt to these requirements efficiently.

Energy Monitoring: Beyond immediate adjustments, REMS provides valuable insights into energy usage patterns, helping operators optimize their energy management strategies over time.

The benefits of implementing a REMS are double. Environmentally, it represents a significant step towards sustainability by reducing a building's carbon footprint. Economically, it leads to substantial savings on energy bills. Adopting REMS does come with its set of challenges, though, including initial setup costs. On average, hotels can expect to see energy savings of up to 20-30% annually. When you factor this in, the return on investment (ROI) becomes clear. With typical ROIs seen within three to five years, the system doesn't just pay for itself; it starts saving money. So, while the upfront costs might seem steep, the long-term savings and environmental benefits are compelling. It's a classic case of spending a bit now to save a lot later, both in terms of operational costs and in doing your bit for the planet. And in today's world, where guests are increasingly eco-conscious, being able to advertise your hotel as energy-efficient can also give you a competitive edge.

Constant vs Variable Temperature Control in Heating Systems

Understanding the difference between constant and variable temperature in a heating circuit is crucial for a multitude of reasons, especially when it comes to operating, and optimizing heating systems across various applications. This knowledge significantly impacts, costs, and overall system effectiveness. Let me guide you through the core reasons why grasping these concepts is essential. Firstly, the decision to use either a constant or variable temperature system can dramatically affect energy usage but it depends on the heating/cooling system design. Variable temperature systems, in particular, stand out for their ability to adjust heating output based on real-time demand. Selecting a temperature control strategy that aligns with the specific needs of a facility or process, it's possible to avoid unnecessary energy consumption. This, in turn, translates into significant reductions in operational costs over time. Optimizing your heating system to match the precise requirements of your application is not just about improving efficiency – it's also about economic sensibility. Comfort and productivity are also at stake when choosing between constant and variable temperature controls. In commercial settings, the ability of variable temperature systems to tailor the indoor climate based on occupancy and the time of day can significantly enhance comfort levels. This personalized control can lead to improved living and working conditions, which, in turn, may boost productivity. Being able to distinguish constant and variable temperature control in your system is the base to understand the heating system design and operations. It empowers us to make informed decisions that effectively balance the demands of comfort, efficiency, cost, and environmental responsibility. As engineers and industry professionals, our ability to navigate these choices can lead to the creation of heating systems that not only meet but exceed our expectations in terms of performance.

Let's explore deeper into how these systems enhance both guests' experience and operational efficiency in a hotel environment. As we know in the context of a hotel, the guests' experience is paramount, and the ambient environment plays a significant role in this regard. The constant temperature (CT) system employed in the hotel lobby serves as the first point of interaction between the hotel and its guests. By maintaining a steady and inviting temperature, the hotel sets a tone of comfort and care from the outset. The AHU system, calibrated to maintain 22°C, ensures that guests are greeted with a consistent environment, regardless of external weather conditions. The function of the AHU, with its strategically placed temperature sensors, demonstrates a blend of technology and guest-centric design; in fact, minimal temperature fluctuations contributes to a seamless and comfortable experience for guests. Moving beyond the public spaces to the private sanctuaries of guest rooms, the variable temperature (VT) systems come into play, offering a personalized experience. Each room becomes a micro-environment, adjustable to the individual preferences of the guest or the occupancy status of the room. This adaptability is central to modern hospitality, where guest satisfaction is closely linked to the ability to customize their stay. The technical integration of the VT systems, with occupancy and/or external temperature sensors, allows for an intelligent response to real-time needs. When a room is unoccupied, the system can lower the heating to a minimum level like 16°C, to save energy. Upon detecting a guest's return or based on a pre-set schedule, the temperature is quickly brought back to a more comfortable level, such as 20°C, or the guest's preferred setting. Continuing our discussion on the practical applications, let's look into the specific use of constant temperature circuits in radiating panels, trench heating etc... These systems, often used for floor, wall, windows heating, provide a unique method of distributing heat throughout a space. The rationale behind employing a constant temperature in these systems is quite insightful. Because the systems heat up large surfaces, they require a sustained and consistent temperature to ensure efficient heat transfer. This approach mitigates the risk of the system becoming inefficient due to the time it takes for the temperature to reach the desired level. Moreover, external factors such as sudden changes in outdoor temperature or the cooling effect of draughts can impact the heating process. If the system were to operate on a variable temperature basis, these external factors could lead to frequent adjustments, potentially reducing the system's overall efficiency and effectiveness. Radiators, instead, are commonly managed through variable temperature circuits, offering a flexible and responsive approach to heating. The device at the heart of this system is known as a variable temperature valve, or VT valve. This valve plays a critical role in adjusting the flow of hot water through the radiators, ensuring that the temperature within the space can adapt to changes based on external factors. The operational logic behind a VT valve is both sophisticated and practical. It continuously monitors the sensors, such as external temperature, and adjusts the flow rate of hot water to the radiators to ensure the indoor environment remains comfortable. For instance, if the external temperature were to drop to -3°C, the VT valve would respond by increasing the water temperature to the radiator, possibly up to 75°C, to compensate for the loss of heat. Conversely, as the external temperature rises, the valve would reduce the outlet flow temperature accordingly, preventing the space from becoming overheated and maintaining a steady, comfortable environment. This dynamic adjustment is crucial for energy efficiency and comfort. It's now clear that by linking the radiator temperature to the external climate, the system ensures that energy is not wasted by overheating rooms when it's unnecessary. Furthermore, this responsive approach allows for a more consistent indoor temperature, enhancing occupant comfort. The use of VT valves in radiator systems represents a smart approach to heating control. It recognises the variable nature of heating requirements throughout the day and across different seasons. This adaptability is particularly beneficial in climates with significant temperature variations, where the efficiency of heating systems can be affected by external temperatures or occupancy. Here's an example of CT/CV valve:

This actuator, utilising an electric motor, operates the valve in either direction, making it more 'open' or more 'closed'. Essentially this is a 3-port valve on the boiler flow. The primary circuit water is strategically directed, either towards the heating circuit, where it aids in maintaining the desired constant/variable temperature, or towards a bypass route that channels it back to the boiler return. This proportional diversion ensures optimal efficiency and precise temperature control within the system.

Thermal comfort solutions

A combi boiler is quite efficient, as it functions as both a water heater and central heating system within a single, compact unit. By heating water directly from the mains when a tap is opened, this innovative system eliminates the need for a bulky water storage tank. The compact design of combi boilers makes them ideal for small demand or where there is limited storage space. Available in both gas and electric models, these boilers predominantly rely on natural gas or LPG for their operation.

On the other hand, a conventional boiler, also referred to as a regular boiler is connected to a separate hot water cylinder and cold-water storage tank. In this system, the cylinder provides hot water for the central heating system, while the boiler receives cold water from the tank to heat for everyday use in taps and showers. Conventional boilers are typically more suitable for larger buildings where the time required to reheat the hot water cylinder can be calculated. A conventional boiler lights up the gas or fuel and the water circulates through the heat exchanger, warming up the water which is transported to the hot water storage tank for easy access. When the ideal temperature is set, the hot water travels through the radiators and outlets to heat up your rooms and to provide hot water for taps and showers.

OIL HEATING BOILER

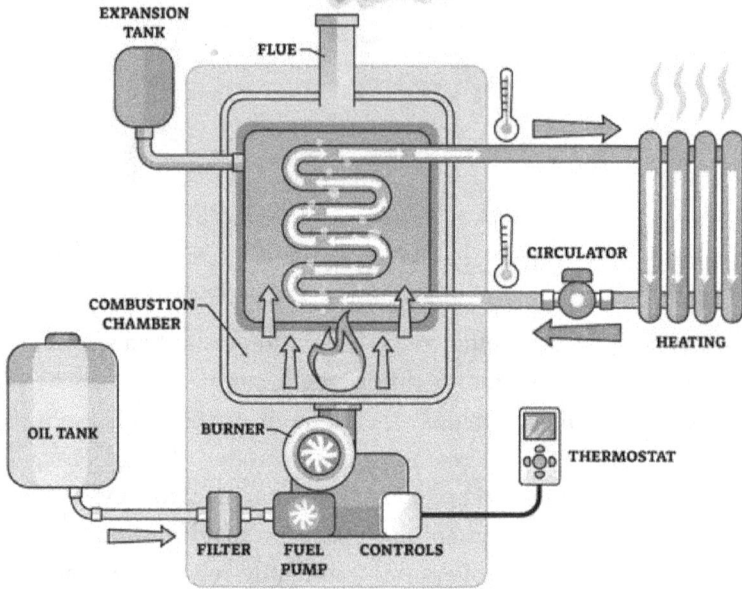

The fundamental difference between a combi boiler and a conventional boiler lies in their individual designs: a combi boiler supplies both heat and hot water through a single unit, while a conventional boiler operates with a separate hot water cylinder and cold-water storage tank. Let's analyse how cylinders work and why they are an important piece of our building equipment.

Direct Hot Water Cylinders

A direct hot water cylinder generates the heat sources; electric immersion heaters are a classic example; in fact, they are located within the cylinder itself to warm the water. These plumbing heating systems are typically found in buildings without access to gas. Some direct hot water cylinders boast dual immersion heaters, enabling users to manage on/off-peak energy tariffs, which in turn, reduces overall operational expenses.

Indirect Hot Water Cylinders

Indirect hot water cylinders instead employs an external energy source – like a traditional central heating boiler or a solar thermal system to heat the water. This energy is channelled into the cylinder via a coil or heat exchanger, which then "indirectly" warms the water. Many indirect hot water cylinders feature an electric immersion heater as a backup, ensuring a reliable supply of hot water in case the primary energy source fails or in hybrid systems to contribute heating based on the demand. Both direct and indirect hot water cylinders can be integrated into vented and unvented systems, providing flexibility to accommodate a variety of project requirements.

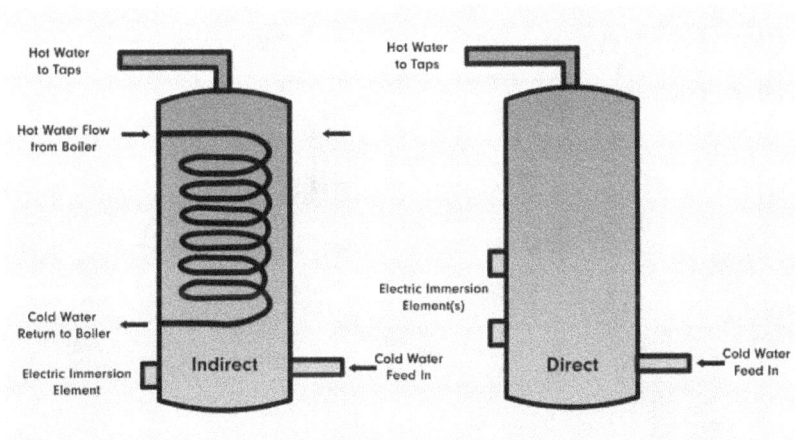

Image source: iHeat

A condensing boiler uses a technology to waste less energy, they are typically at least 93% energy efficient, rising up to 97% for the very best models on the market. The first thing to remember about conventional boilers and condensing boilers is the heat input from the burner is the same (roughly 250-300 degrees Celsius). As mentioned above, a conventional boiler or standard boiler has one heat exchanger, water enters the heat exchanger at the coldest (and generally lowest) point and collects heat, leaving the heat exchanger and flowing off to the radiators and hot water tanks. The flue gases temperature can be anything up to 250 degrees Celsius and that heat is expelled into the atmosphere and never used again, which means it is wasted. A condensing boiler instead has two heat exchangers. The water enters the secondary heat exchanger chamber, picking up latent heat from the hot flue gases, the moisture in the gases condenses into drops which are expelled into the drain. From there the water enters the primary heat exchanger to collect more heat before flowing to the radiators and hot water tanks. In this case the flue gases temperature will be around 55 degrees Celsius which means up to 200 degrees of heat is being transferred to your heating water before the main exchanger, significantly increasing the efficiency.

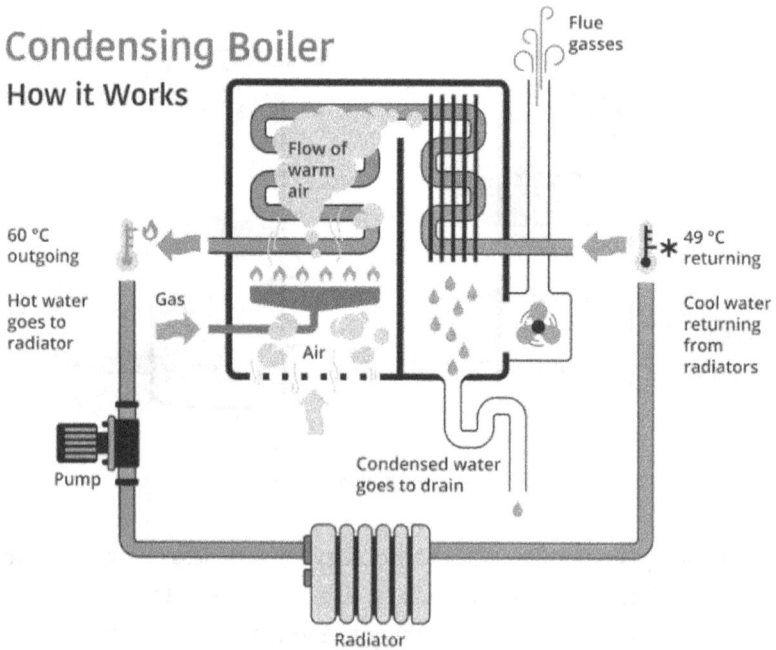

Condensing Boiler
How it Works

Flue gasses

Flow of warm air

60 °C outgoing

49 °C returning

Hot water goes to radiator

Gas

Cool water returning from radiators

Air

Pump

Condensed water goes to drain

Radiator

Image source: boilerhut

Gas Tank-Type Water Heater:

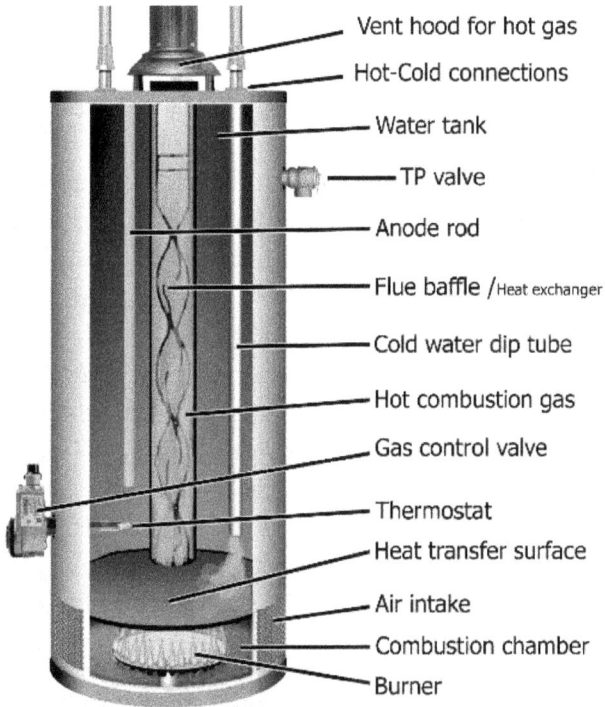

Vent hood for hot gas

Hot-Cold connections

Water tank

TP valve

Anode rod

Flue baffle /Heat exchanger

Cold water dip tube

Hot combustion gas

Gas control valve

Thermostat

Heat transfer surface

Air intake

Combustion chamber

Burner

© Gene Haynes

A gas-powered water heater has cold water brought into the tank through a dip pipe. This water is heated with a gas burner which releases extremely hot and toxic air through a chimney in the middle of the water heater tank. The chimney moves the hot air outside while heating the metal of the chimney and as it heats up, the surrounding water is heated as well to maximize the efficiency. As the water becomes warmer it rises in the tank where it is drawn off by the hot water discharge pipe to provide hot water wherever it is called for. The hot water discharge pipe is much shorter than the dip cold supply pipe as its goal is to funnel off the hottest water, which is found at the very top of the tank. The gas burner that heats the water is controlled by a gas regulator valve mounted on the side of the water heater, which includes a thermostat that measures the temperature of the water inside the tank and turns the burner on and off as needed to maintain the set temperature of the water.

The Tank

The tank of a water heater consists of a steel outer jacket that encloses a pressure-tested water storage tank. This inner tank is made of high-quality steel with a vitreous glass or plastic layer bonded to the inside surface to prevent rusting. In the centre of the tank is a hollow exhaust flue through which exhaust gases from the burner flow up to an exhaust vent. In most designs, a spiral metal baffle inside the flue captures heat from the exhaust gases and transmits it to the surrounding tank. Between the inner storage tank and the outer tank jacket there is a layer of insulation designed to reduce heat loss. You can also supplement the insulation by adding a fibreglass insulation tank jacket to the outside of the hot water heater. These are inexpensive and easy to install, but it is important to avoid blocking the burner access panel and the flue hat at the top of the tank.

Inside the Tank

In the tank, there will be a metal rod (magnesium or aluminium), called a sacrificial anode. The anode rod is bolted and fastened to the top of the tank and extends deep into the tank. Its purpose is to draw rust causing ions in the water to itself, preventing the metal tank from corroding. Some models do not have a separate anode rod but instead have a hot water outlet pipe that is coated with magnesium or aluminium to serve the function of an anode. If hot water coming from faucets becomes smelly or discoloured, it may be an indication that the anode rod has been consumed. Cold water is provided to the tank by a cold-water supply line controlled by a shutoff valve. It is important to know where the water supply shutoff valve is located so you can close it when maintenance is required. Shutting off the cold-water supply effectively shuts off the water flow entirely, since it is the pressure from the cold water entering the tank that keeps the hot water flowing outward. In many installations, the cold-water supply shutoff valve will be identified by a blue handle.

Temperature and Pressure-Relief Valve

Another key safety feature of a hot water heater is the temperature and pressure-relief (T & P) valve. The purpose of this valve is to relieve excessive temperature or pressure build-up inside the tank if it approaches the limit of the tank's design. This valve is located on top of the tank and often is threaded directly into the tank top itself. To test the valve, lift up on the handle slightly; tank water should discharge out of the overflow pipe. If it doesn't operate properly, the T & P valve should be replaced.

Tank Drain Valve

The hot water tank can build up sediments in the bottom of the tank over time, leading to several problems. A water heater full of sediments will not heat efficiently, and you may hear bubbling, gurgling sounds caused by the moisture-saturated sediments boiling. By periodically draining the tank using the tank drain valve, these sediments are removed, and problems are avoided.

Some heating or cooling systems work fine even without proper water treatment, while others, even with careful maintenance, get damaged by corrosion and fail. Why do some systems last despite neglect, while others break down despite careful upkeep? The truth lies beneath the surface, in the interaction between water, metal, and the myriad conditions they're subjected to. Corrosion, is an insidious enemy for most common metals, and it does not act randomly. Its pace and the risk of pipework failure are influenced by a multitude of factors – from the chemical and microbiological composition of the environment to the physical conditions such as temperature, flow rate, and crucially, the thickness of the metal involved. Water is brilliant for moving heat around in buildings because it can hold a lot of heat, works well over a range of temperatures, and is safe and easy to get hold of. But, its ability to carry electric charge means it can cause rust in metal pipes and parts. While using plastic pipes might seem like the perfect solution, it comes with its own set of problems. In metal systems, oxygen, which speeds up rusting, is used up quickly, not really causing much harm. But, if you use plastic pipes, oxygen stays around longer, making any remaining metal parts rust faster. This highlights why treating the water in all heating and cooling systems is crucial, especially when plastic pipes are involved. The need for water treatment becomes even more critical to prevent quicker corrosion of any metal parts left in these systems. The importance of water treatment in the management of building services systems is immense. It acts as a crucial line of defence against the widespread issue of corrosion. If not managed properly, corrosion can seriously harm the structural strength and performance of heating and cooling systems. So, what's involved in water treatment, and why is it so important? At its heart, water treatment adjusts the water's qualities to better suit the systems it's used in. This involves managing the elements that lead to corrosion and the build-up of scale, transforming potentially harmful water into a safe carrier for managing temperatures within our buildings. The process of water treatment starts with understanding the chemical and biological composition of the water. Water from various sources might contain different contaminants and microorganisms, each posing a risk of starting or speeding up corrosion. A familiar example is the high levels of dissolved oxygen that can react with metals in the pipes, causing them to wear thin over time and potentially lead to leaks or failures. To tackle this, water treatment plans often include corrosion inhibitors. These special chemicals, when added to water, can either coat the metal surfaces, protecting them from corrosive elements, or neutralize the corrosive agents in the water. The aim is to put up a barrier that keeps the metal and the elements causing corrosion apart, slowing down or even stopping the corrosion process altogether. Moreover, the role of biocides and cleaning agents is critical in keeping the water quality up to the mark. Biocides help in managing the growth of microorganisms that can form damaging biofilms on the pipes' inner surfaces. Not only do these biofilms speed up corrosion, but, as we know, they can also harbour bacteria that pose health risks. A full cleaning of the systems before they're put into use, often involving biocides and chemical cleaning (for example the famous "chlorination"), is crucial to start their operational life on the right note, free from contaminants that could encourage corrosion. However, water treatment isn't a fix-all solution on its own. Its effectiveness is closely linked to the design and functioning of the system. For example, a system with areas where water can stagnate, or that allows fresh, oxygen-rich water to enter, can undermine even the best water treatment efforts. Therefore, it's essential that systems are designed first with features that minimize these risks and monitored later by the maintenance team, allowing water treatment chemicals to work effectively; in fact, water treatment is an ongoing process. Simply treating the water isn't enough. Regular monitoring of water quality, along with necessary adjustments to the treatment process, is vital. This ensures that the system can adapt to any changes in water chemistry, modifications in the system, or shifts in how it's operated. This proactive approach is what ultimately ensures the long-term efficiency and durability of heating and cooling systems. Corrosion is when materials break down because of chemical reactions with their surroundings. It's really important to understand how corrosion works so we can try to stop it from happening.

How Corrosion Happens

Corrosion mainly occurs through a process that involves electricity at a tiny scale, where electrons move from the metal to oxygen in the water. Water helps this process along, acting like a bridge that allows the electrons to move, which eventually causes the metal to weaken and break down. Sometimes corrosion affects a metal surface evenly, which can be bad but is usually something we can deal with. Other times, it attacks specific spots, causing things like pitting, cracks, or openings. This type of corrosion is much sneakier and can cause serious damage before you even realize it's happening. Pitting corrosion is a big problem for building systems. It starts in tiny areas where the metal becomes more reactive, and these spots get eaten away, forming pits. This is especially tricky because it can happen fast and without obvious signs until it's quite advanced. The water's properties, like its chemistry and temperature, can make this worse. Oxygen has a mixed role in corrosion. At first, it speeds things up by grabbing electrons from the metal. But if a system is closed off and runs out of oxygen, things can slow down. However, if new oxygen gets in, it's like giving corrosion a new lease on life. Then there's the issue of tiny life forms, like bacteria, that can make corrosion even worse. They form slimy layers called biofilms on metal surfaces and can create conditions that really speed up corrosion. These biofilms are tough to get rid of and often require special treatments to keep them in check. Understanding these corrosion mechanisms is key to keeping building systems safe and working well. When heating and cooling systems fail due to corrosion, it's not an overnight event but a gradual process. It starts with the initial rust forming on metal parts, then moves to more serious damage like deep pits or cracks, and eventually, the system's parts might break or leak. This journey from small rust spots to major failures can cause systems to work poorly, leak, or even stop working altogether. Preventing such breakdowns is all about knowing exactly how corrosion works and then taking steps early to stop it. Engineers have to think ahead, using the right materials and designs to keep systems running smoothly and safely for as long as possible. To stop corrosion from happening, there's quite a bit to think about:

Choosing the Right Materials: Using metals that resist rust can keep systems safe but might cost more at the start.

Smart Design: Systems need to be set up so water moves freely without staying still in any spot, which could lead to rust or bacteria growth. Also, making sure there's a way to check and maintain the system easily is key.

Keeping Oxygen Out: In systems that are closed off, it's important to stop oxygen from getting in because it can cause rust. Special designs or equipment can help remove oxygen from the water.

Staying on Top of Water Quality: By regularly checking the water for signs of rust or bacteria and making sure the protective chemicals in the water are at the right levels, you can catch problems before they get worse.

Checking the System: Regularly looking over the system and using tests that don't damage it can help spot issues early.

Keeping up with Maintenance: Making sure the system is clean and that any equipment used to add chemicals to the water is working right is crucial for keeping everything running smoothly. It's also important for everyone involved with the system, from designers to maintenance team, to understand how corrosion works and what they can do to prevent it. Keeping good records of what's been done to the system can also help with troubleshooting and planning for the future.

As mentioned, for ensuring the longevity and efficiency of heating and cooling systems, there's a clear need to focus on maintenance and the strategic selection of materials and technologies designed to combat corrosion. Firstly, setting up a detailed schedule for checking and maintaining the system is crucial. This includes routine checks on water quality, looking over the system regularly, and making sure it's clean. It's not just about ticking boxes; these activities are tailored to what each specific system needs to stay running smoothly. Incorporating modern technology plays a big role too. For example, using online sensors and systems that automatically adjust the levels of treatment chemicals based on real-time data can make a big difference. These tools help keep the water treatment precise and take some of the burdens off maintenance staff, making the whole process more efficient. Training for the staff is another important piece of the puzzle. It's essential that everyone involved in the maintenance of these systems knows how important their role is in preventing corrosion and keeping everything running smoothly. Sometimes, bringing in experts in water treatment can provide valuable insights and help improving the maintenance strategy. Being ready to act quickly when issues pop up is also key. Whether it's adjusting the chemicals used in water treatment, fixing leaks to stop air from getting in, or doing extra cleaning, quick responses can prevent small problems from turning into big ones.

Electric Water Heater

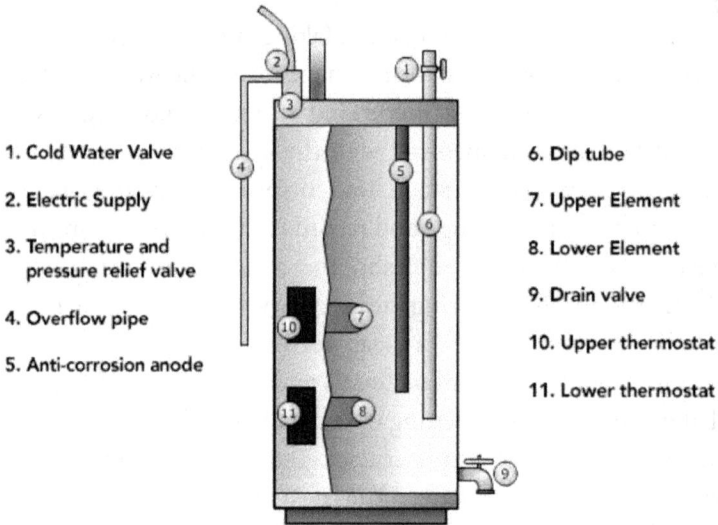

1. Cold Water Valve

2. Electric Supply

3. Temperature and pressure relief valve

4. Overflow pipe

5. Anti-corrosion anode

6. Dip tube

7. Upper Element

8. Lower Element

9. Drain valve

10. Upper thermostat

11. Lower thermostat

An electric water heater works essentially the same way as a gas water heater. It brings cold water in through the dip pipe and heats it using the electric heating elements inside of the tank. The hot water rises in the tank and is moved throughout the building through the hot water pipe. As with the gas water heater, an electric water heater has a thermostat, temperature and pressure relief valve, a drain valve, the tank is insulated, and it has an anode rod. The only big difference is the water is heated by electric elements which means power supply is needed.

Modular Boilers

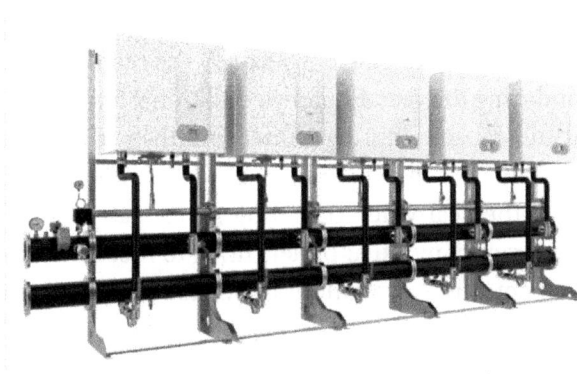

Modular boilers are designed as an alternative to large single boilers and offer a very efficient approach to commercial heating. A modular boiler system can provide hot water or steam for room heating and hot water. Smaller modular boilers are physically easier to transport and handle than very large boilers. For retrofit projects they can be installed alongside the existing boiler as they are capable of high outputs from a compact size. This enables a smooth changeover from the old heating system to new with little or no system downtime. This makes the modular boiler system very popular in schools, the health sector and hotels where continuous heating is critical to their business. Often modular boiler ranges have lightweight designs and will fit through a single doorway, making them ideal for rooftop or basement plant rooms and high-rise residential buildings, they operate together in parallel to provide variable amounts of heat (via water or steam) to best match system load. Each modular boiler provides a percentage of the system's heating capacity, and additional units can be added to increase capacity. A programmed controller coordinates the number of boilers needed to meet real-time facility demands. This allows each modular boiler to be switched on only when needed (minimizing run time) and operating more efficiently. In contrast, conventional boilers are often sized to satisfy large loads (e.g. whole buildings in cold weather), and when facility demands do not require the full heating capacity, efficiency is sacrificed as the larger conventional boiler is throttled down. Firing boilers independently allows a modular boiler system to efficiently satisfy the building heating load fluctuations during the year. Equally, a traditional boiler may typically run at near-full capacity during winter months with a large heating load requirement, but then operate at a consistently lower efficiency for the rest of the year when operating under part-load conditions. Below is a diagram of modular boilers firing to meet different heating load requirements due to variations of outside air temperature.

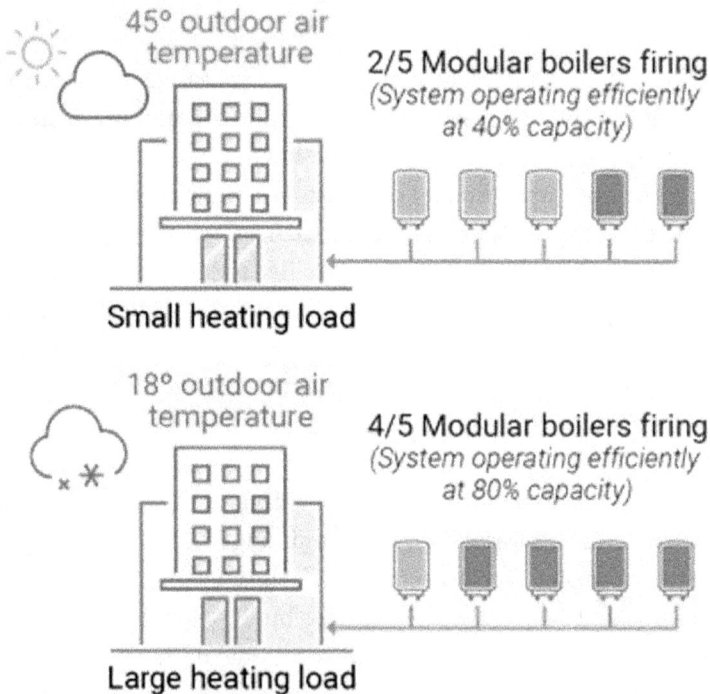

45° outdoor air temperature

2/5 Modular boilers firing
(System operating efficiently at 40% capacity)

Small heating load

18° outdoor air temperature

4/5 Modular boilers firing
(System operating efficiently at 80% capacity)

Large heating load

Two or more modular boilers are connected to meet a fluctuating or partial heating demand. They can be either conventional and used in high-temperature systems, condensing and used in low-temperature systems or mixed (e.g. two conventional, two condensing) in a hybrid system, depending on design needs. A system controller monitors the temperature of the water circulating through the system, outside air temperature, and building heating demand and fires boiler modules just enough to meet the demand or turns off modules as demand decreases. A crucial benefit of modular boilers is that technicians can perform required maintenance activities at any time, keeping the system operating. This same feature improves overall system resilience, allowing the building or process to continue operating despite a failure. As already stated, a modular boiler system is a collection of boilers that work together and each module can be a separate boiler capable of running independently of the others, they can be installed alongside one another in a horizontal arrangement or as a vertical stack of boiler modules one above another, but the real differentiating component that allows modular boilers to function properly as a collection of boilers is the system controller. The controller for a multiple-boiler system enables the modulating boilers to be operated such that they modulate together to match the load. The controller can optimize the system operation by firing the optimal number of boilers needed to match the load with each operating boilers at its most efficient state. A stand-alone controller can operate the boilers using its internal programme while an integrated controller into a building automation system (BAS) receives the required operating temperature control point from the BAS.

Modular boiler system controllers can be configured to sequence boilers in either cascade mode or unison mode.

- Cascade control initiates a boiler operation at its lowest heating rate, modulates it to its maximum rate, and then starts the next boiler to meet the system load. This method operates the minimum number of boilers to satisfy the load requirements. Cascade control is used for systems where there is no significant advantage to operating boilers at part load. This strategy is often applied to non-condensing boilers, providing constant-temperature hot water or constant-pressure steam.

- Unison control takes advantage of the higher part-load efficiencies available at low boiler heating rates. Unison control initiates boiler operation at its lowest rate. If the load requires additional heat, the controller enables additional boilers at their lowest rate and then, if needed, modulates the boilers to higher rates to match the system requirements. The unison control strategy is well suited for operating condensing boilers in variable-temperature systems or systems with variable loads. In both cases, boiler control can include a lead/lag control status.

- Lead/lag control of modular boilers involves the designation of a single boiler to serve as the starting boiler upon a call for heat – this is the lead boiler. A second (lag) boiler, as well as additional boilers, will be enabled as required. The boiler controller will then rotate the designation of lead/lag status based on operating hours. From an engineering and costs management perspective, proper lead/lag control results in not only efficient system operation, but also equal wear, less cycling, and less stress on the boiler and system components improving the boiler's lifespan.

Unison Control

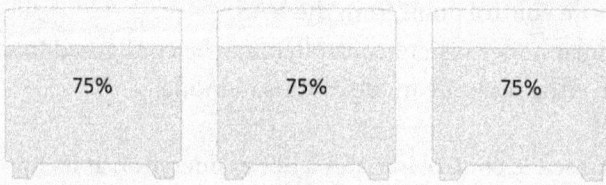

| 75% | 75% | 75% |

Cascade Control

| 100% | 100% | 25% |

Have you ever walked into a plant room and been amazed at the complex array of equipment that keeps our buildings warm in the winter and cool in the summer? I want to share with you a fascinating piece of technology that might already be there, or could be a key part of your future installations. It's the Microfill pressurisation unit, a system that defines simplicity and efficiency. While conventional pressurization units use pumps to top-up the pressure in the system, the secret to the Microfill's exceptional performance lies in its direct connection to the main cold-water supply. This approach eliminates the need for a traditional pump, which is a significant leap forward in pressurization technology.

Dimensions

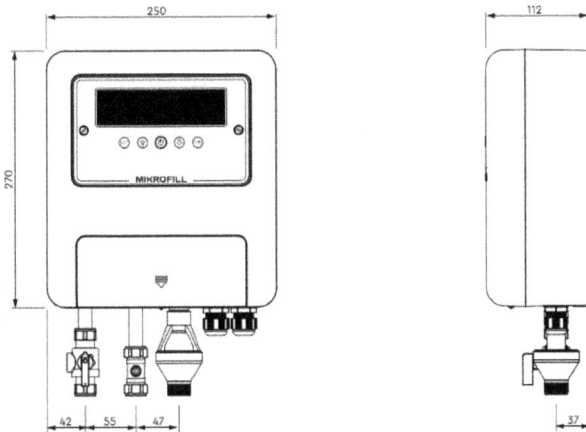

Some technical features:

WRAS Approval: Ensures compliance with high standards of quality and safety.

Electrical Efficiency: With a supply voltage of 230 Volt and a full load current of just 1 Amp, the unit is exceptionally energy-efficient.

Optimized Design: Weighing only 5kg and with a nominal flow rate of 14 l/min at 2 Bar, it's both compact and powerful.

Installation Flexibility: The unit's design allows for easy integration into existing systems, with 15mm compression for both cold water inlet and system outlet.

Integration: The unit also incorporates two BMS relays offering volt free contacts for the remote indication of low- or high-pressure conditions within the system.

When I first came across this, during a company CPD training, I found out the ability of this system that connects directly to the main cold-water supply, eliminating the need for a conventional pump, and I thought: This is a game-changer.

Simplicity in design: By connecting directly to the mains water supply, Microfill simplifies the entire pressurization process. It's like plugging in a device and having it work impeccably, without the intricacies and potential failures of a separate pump. This direct connection is not just about straightforwardness; it's also about efficiency. The system only uses what it needs from the mains supply, ensuring that no energy is wasted on pumping actions that aren't required.

Reliability: With fewer moving parts and less mechanical complexity, the risk of breakdowns is significantly reduced. This reliability is something we all seek in our busy lives, where a heating or cooling system failure can be a major inconvenience in any organization you work. "Peace of mind."

Environmental responsibility: In an era where every bit of energy saving counts towards environmental conservation, the Microfill stands out. It aligns perfectly with our growing need for sustainable solutions.

For those, like me, who are curious and involved in facility management and technical tasks, recognizing a Microfill unit in your plant room can be enlightening. It's a compact, efficient unit, marked by its direct connection to the mains supply. Knowing that this technology is part of your system can be reassuring – a sign that your facility is on the cutting edge of efficiency and sustainability. Exploring and sharing innovation technologies like this, is a source of immense fascination. It's a reminder that innovation is not always about adding more; sometimes, it's about smartly taking away the unnecessary stuff.

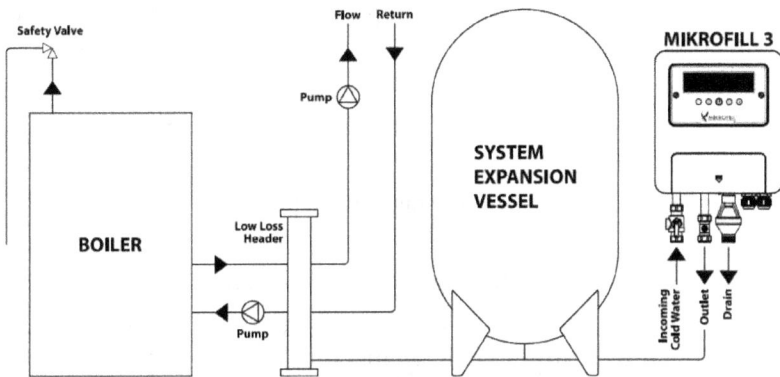

DIAGRAM OF A TYPICAL INSTALLATION

Image source: Mikrofill inspired efficiency

Biomass Boiler

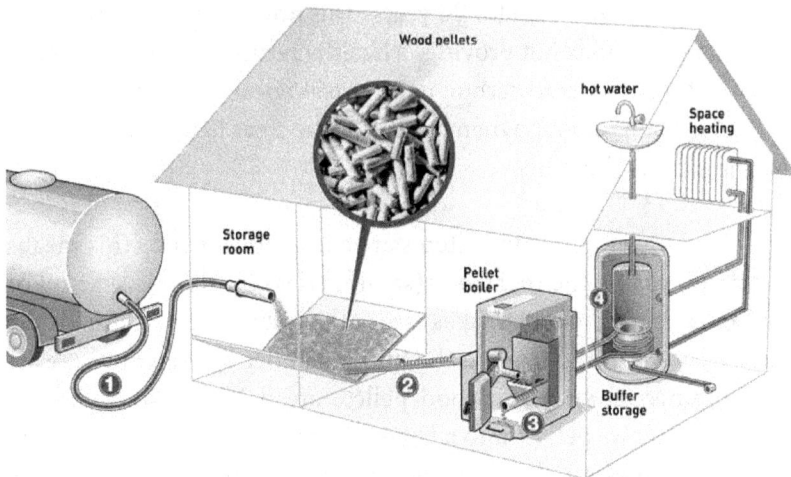

The word biomass covers a whole range of biological mass. When it comes to heating, in most of the cases wood is the type of biomass used as the fuel in the boilers. There are three main types of fuel: wood pellet, wood chip and logs; however, there are some lesser used types on non-woody biomass like miscanthus and cereals. The integration of a biomass system is relatively simple – in most cases the systems are retro fitted and replace a fossil fuel boiler. The biomass boiler replaces the existing fossil fuel boiler, but the downstream heating system will remain the same, so there is no need to replace the radiators or whatever emitters are in place.

There are three main advantages in switching to biomass:

- Fuel savings – Heating with biomass can be a lot cheaper. Where there is an existing wood supply, you can see savings of around 80% against oil after factoring in costs for processing the wood
- Using biomass can qualify you for the government's Renewable Heat Incentive (RHI). The heat used is recorded on heat meters and quarterly readings should be submitted to allow owners (commercial systems) to receive a compensation from the OFGEM which pays every quarter for the heat used for 20 years. This not only pays for the initial capex but provides a healthy return over the lifetime.
- It is considered carbon neutral as it releases the same amount of CO_2 when burned as the trees have absorbed during their life.

Wood pellet and chip systems are automatically fed. This means that if you ensure the fuel stock (same as if you would have an oil or LPG tank), then the boiler takes care of the rest where it automatically takes wood from the store to the burn chamber. On larger domestic and commercial systems, wood pellets are delivered in bulk by a tanker that blows pellets into a large fuel store. Pellets are transferred from the store to the boiler either via an auger or a suction system. Biomass boilers use a process called gasification to produce heat.

There is a two-stage burn process where the fuel is initially burnt at temperatures of around 600°C which releases gases. These gases are then re-burned which allows the temperature in the burn chamber to reach around 1200°C. The hot gases pass through a metal heat exchanger, which then heats the water on the other side.

Thermal stores / Hot water storages are used to be more efficient, large tanks of water (upwards of 30 litres per kW of boiler output) that act as a heat battery by storing the hot water from the boiler which is then circulated to the central heating system. The biomass boiler has temperature sensors in the thermal store, so it is responsible for keeping it at predefined temperatures. Here the boiler will automatically switch itself on when the thermal store drops below a certain temperature, then when the thermal store is up to temperature it switches itself off to conserve fuel. The thermal store has a main system pump after it which circulates the warm water to the building. Each end user point will be able to send a signal back to the biomass plant room when there is a demand for heat, which results in the pump circulating the hot water.

A pivotal component that often goes unnoticed but plays a crucial role in the optimization of these systems is the Low Loss Header (LLH). I want to explain the significance of the LLH, demonstrating how it facilitates the creation of primary and secondary circuits, thus enhancing water flow balancing and extending the lifespan of boilers. At the heart of a typical heating system there is the boiler, whose efficiency is greatly influenced by the heat exchanger. For the heat exchanger to operate at peak efficiency, it is imperative that the water speed passing through it is maintained within specific parameters. However, in many heating installations, the flow rate through the system may either exceed or fall short of the boiler's optimal flow rate. This discrepancy can lead to inefficiencies and potential damage to the system. The installation of a Low Loss Header into the system addresses these challenges by establishing a primary circuit. This circuit acts as a buffer, maintaining water pressure at the required rate, regardless of fluctuations in the heating system's demand. Such a mechanism ensures that the heat exchanger functions efficiently, thereby safeguarding the boiler's longevity. Moreover, water temperature plays a critical role in the performance of a heating system. Two primary issues arise from temperature discrepancies: thermal shock and inefficiency in condensing boilers, for example. Thermal shock occurs when there is a significant temperature difference between the flow and return, potentially shortening the lifespan of the heat exchanger. On the other hand, for condensing boilers to achieve maximum efficiency, they must operate within certain temperature parameters. Ideally, the return temperature should not exceed approximately 55°C to enable the boiler to enter condensing mode, where it operates most efficiently. A Low Loss Header, equipped with temperature controls, can cleverly manage these temperature differences, preventing thermal shock and facilitating optimal conditions for condensing boilers.

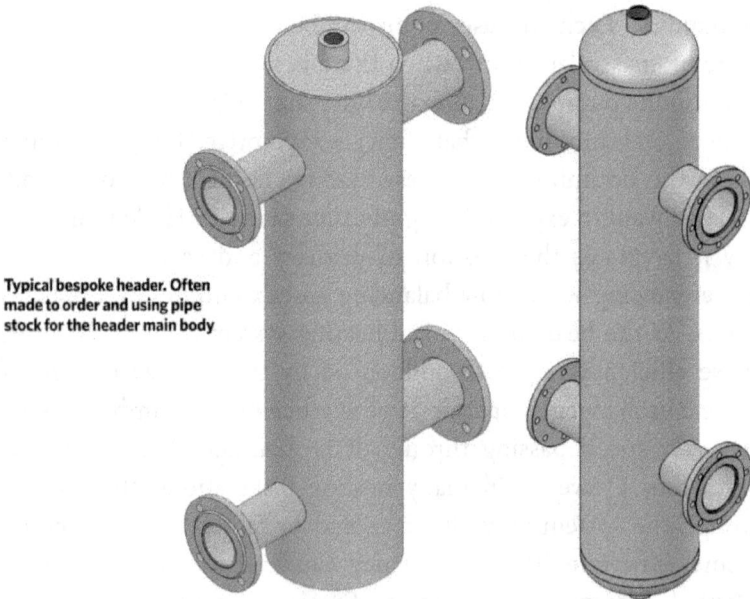

Typical bespoke header. Often made to order and using pipe stock for the header main body

Naturally, you will see low loss headers in commercial installations. In domestic heating systems, they can be installed where the internal boiler pump doesn't have enough power or speed for the whole system.

What's the path of least resistance?

Water, like electricity, follows the path of least resistance. This means that given multiple pathways, water will choose the route that requires the least amount of energy to travel. This path is typically the one that is most direct and involves the least friction or obstruction, which often aligns with the shortest physical distance between two points. The water will trend towards the most efficient route that gravity dictates. The LLH exploits these fluid dynamics; in fact, the chamber inside a low loss header creates a shortcut across the flow and return pipework. If a boiler with an internal pump, pumps to a low loss header, almost 100% of the water will return back to the boiler and very little flow, if any, will continue on to the system. This hydraulic separation allows a system pump to be installed on the other side of the header and operate in much the same way with minimal disruption to the boiler side of the Header. Here is an example of an LLH in a heating system which includes domestic hot water, radiators and underfloor heating.

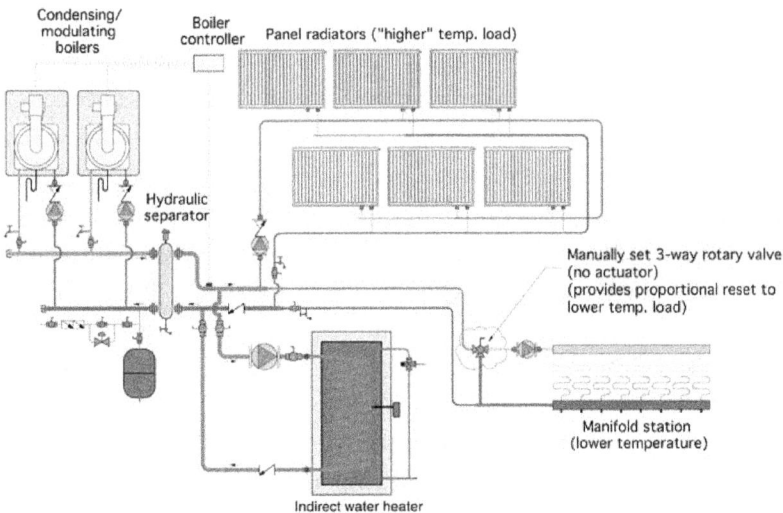

Condensing/modulating boilers

Boiler controller

Panel radiators ("higher" temp. load)

Hydraulic separator

Manually set 3-way rotary valve
(no actuator)
(provides proportional reset to lower temp. load)

Manifold station
(lower temperature)

Indirect water heater

As water flows into the wide section of the header, it slows down significantly, to about half the speed it was going in the pipes. This slower flow lets the bits of dirt in the water fall to the bottom of the header, while any air bubbles float up to the top. What's more, the LLH can catch all types of dirt, not just the metal bits that stick to magnets like those caught by magnetic filters. It can also trap non-magnetic materials such as copper, brass, tin. Over time, even bits of steel and iron, which usually stick to magnets, can lose their magnetic pull and will also be caught by the LLH. It's important to mention that this cleaning action works best if the header is installed vertically, rather than horizontally. Some LLHs have a special mesh inside to make the cleaning process even better. Ensuring the low loss header is suitably insulated is crucial, as without it, there's a risk it could become an unintentional radiator due to its size, leading to unwanted heat loss. While the LLH could be an added expense, the necessity for multiple pumps or heat sources leaves little room for alternative solutions that possess the same level of refinement. A potential complication with the incorporation of any form of hydraulic separation, such as a low loss header, is the phenomenon known as 'distortion'. This term describes the scenario where the boiler must operate at higher temperatures to achieve the desired temperature for emitters like radiators or underfloor heating systems. This situation typically arises because the flow rates on either side of the low loss header are not identical. Consequently, these elevated temperatures at the heat source can result in a marginal reduction in efficiency, particularly in gas boilers. The effect is even more pronounced with heat pumps, accompanied by other challenges associated with high-temperature systems. The root of this distortion is the mixing of the flow and return waters within the header; it does not refer to a warming of the boiler return, but instead to uniformly raise the temperatures for the heat source and potentially the emitters. However, it's important to state that these issues should not deter the installation of a low loss header. If a low loss header is considered essential, attention to minimizing distortion during the commissioning phase is key to enhancing efficiency and overall system performance.

Heat Pumps

Heat pumps can source heat from the air, water, or ground and concentrate it for use inside. The most common heat pump water heater is the air-source which extracts ambient heat from the surrounding air and circulates that heat around within the pump to further increase the temperature. There are two types of air source heat pump, air-to-air and air-to-water. The Air-to-air heat pumps, commonly known as air conditioning, provide space heating or cooling by blowing air in the building areas. The Air-to-water heat pumps instead heat radiators, underfloor heating and hot water. These systems are two to three times more efficient than an electric water heater because it moves heat around instead of generating heat by itself. The efficiency of a heat pump water heater relies on the quality of the system, the calibre of the installation, the average temperatures of your climate and the positioning of the compressor unit.

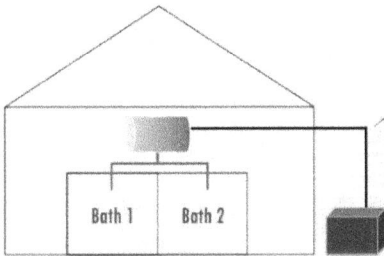

Bath 1 | **Bath 2**

SPLIT SYSTEM
REFRIGERANT TUBE

The longer the refrigerant tube, the lower the efficiency. Heat is rapidly lost through long refrigerant tubes.

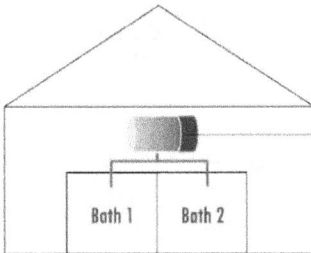

Bath 1 | **Bath 2**

INTEGRATED SYSTEM
NO REFIRGERANT TUBE

The refrigerant of an integrated heat pump runs internally, minimizing heat loss to the environment.

There are two types of heat pump water heaters:

- Split systems: the water tank and compressor are separate – just like an air conditioning system.
- Integrated units: both the compressor and the water tank are placed together.

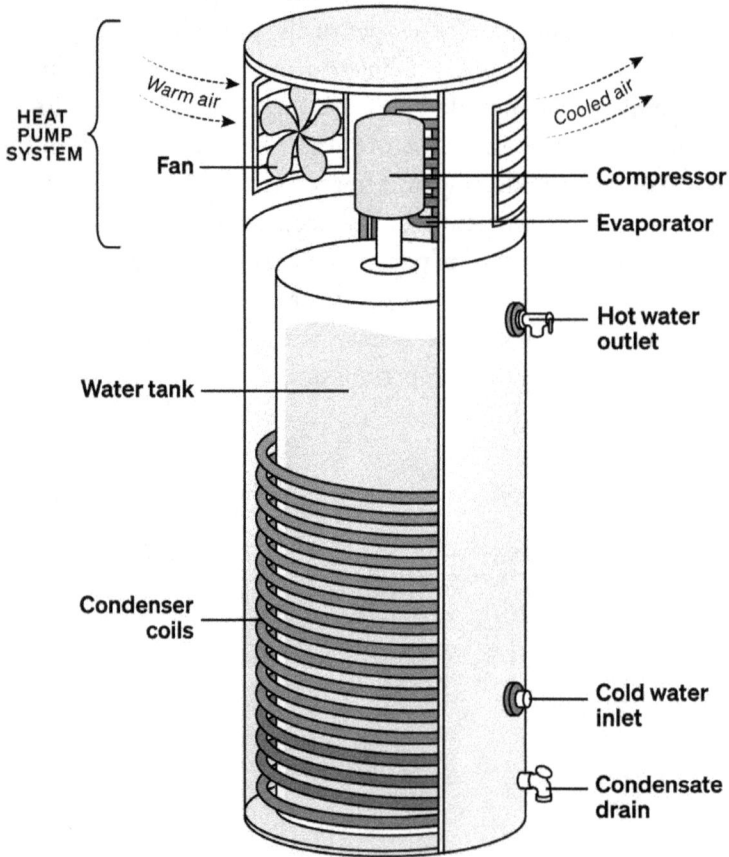

HEAT PUMP SYSTEM

Warm air

Cooled air

Fan

Compressor

Evaporator

Hot water outlet

Water tank

Condenser coils

Cold water inlet

Condensate drain

Image source: Energy star

The first step:

In the process of bringing the ambient air into the unit a fan is hooked up to a compressor within the water heater. The heat pump is the top area (image above) of the unit that houses the compressor, evaporator coils, and expansion valve. The ambient air from outside of the unit is brought into the unit, the air enters the unit and flows through the compressor and evaporator coils so that the heat energy from the air is absorbed into evaporator coils. It's important to note that the air doesn't have to be "hot" as heat pumps are extremely efficient at absorbing heat energy from the air, even at temperatures as low as five degrees Celsius. That said, the higher the air temperature, the more efficient the unit will be.

The Second Step:

Refrigerant is pumped through the evaporator coils and because the refrigerant has a low boiling point (which means it will boil at cooler temperatures than water), the heat from the air being pushed through the system brings the refrigerant to a boiling point becoming gas. When gas is compressed, its overall temperature increases; in fact, the compressor adds pressure onto the gas which continues to increase in temperature.

The Third Step:

The first heat transfer was from the air to the refrigerant. Our second will be from the refrigerant to the water itself. The heated refrigerant is housed in the condenser coils. Water comes in near the bottom of the unit and is then pumped to the top of the unit, where the heat is transferred from the refrigerant in the condenser coils to the water.

Hybrid Variable Refrigerant Flow

HVRF, or Hybrid Variable Refrigerant Flow, is a cousin of the well-known VRF (Variable Refrigerant Flow) air conditioning system. At its core, VRF technology is pretty clever, utilising one external unit connected to multiple internal units. This setup allows for the cooling or heating of different rooms and spaces independently, all thanks to the flow of refrigerant through piping. It's a fantastic solution for large buildings where different areas might have different climate control needs. Now, where HVRF starts to shine is in its innovative twist on the traditional VRF system. Instead of relying solely on refrigerant to transfer heat and cool spaces, HVRF introduces water into the equation. This hybrid approach combines the best of both worlds: the efficiency and flexibility of refrigerant-based systems with the environmental and safety benefits of water-based systems. This innovative system joins the power of a unique two-pipe refrigeration combined with water to deliver best comfort, making it an ideal choice for a wide range of buildings, including offices, hotels, medical centres, schools, high-rise buildings, and shopping centres.

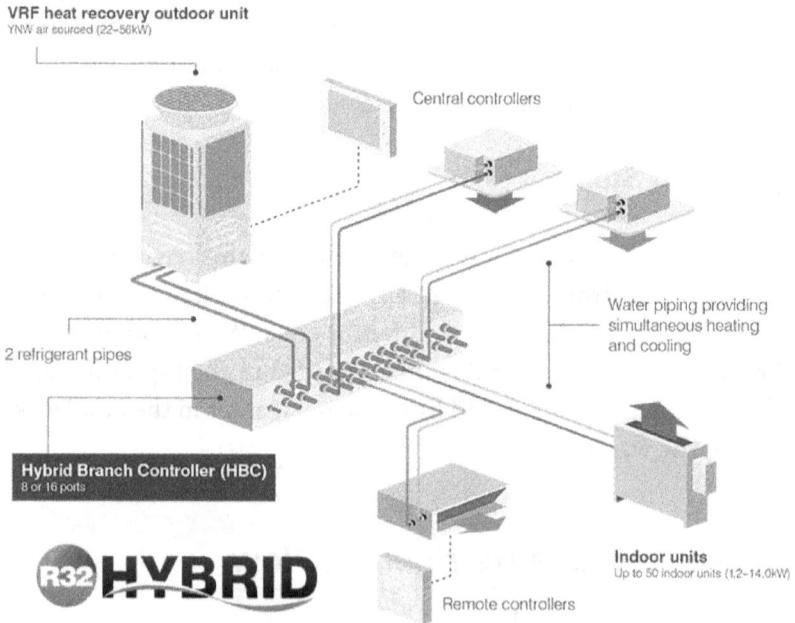

VRF heat recovery outdoor unit
YNW air sourced (22–56kW)

Central controllers

2 refrigerant pipes

Water piping providing simultaneous heating and cooling

Hybrid Branch Controller (HBC)
8 or 16 ports

R32 HYBRID

Remote controllers

Indoor units
Up to 50 indoor units (1.2–14.0kW)

HVRF technology has the ability to operate with approximately 75% less refrigerant compared to traditional Variable Refrigerant Flow (VRF) systems. This substantial reduction is made possible by replacing refrigerant with water in the final stages of energy transfer. Water, being a natural resource with no GWP (Global-warming potential), serves as an excellent medium for transferring heat without contributing to global warming. Another benefit is costs, the traditional refrigerant-based system needs a comprehensive leak detection setup to quickly identify and mitigate leaks, protecting both the environment and building occupants. This equipment represents an additional upfront investment and ongoing operational expense. HVRF systems, by largely containing the refrigerant to the external unit and minimizing its use within the building, significantly reduce the dependency on sophisticated leak detection equipment in occupied spaces. This not only lowers the initial investment in such systems but also decreases the energy consumption and maintenance requirements of running these detection systems. At the heart of the HVRF system is an HBC (Hybrid Branch Controller) box, which is connected to the outdoor unit via traditional refrigerant piping. Between the HBC box and the indoor fan coils, the system uses water piping but still offers high sensible cooling/heating and stable room temperatures for maximum comfort.

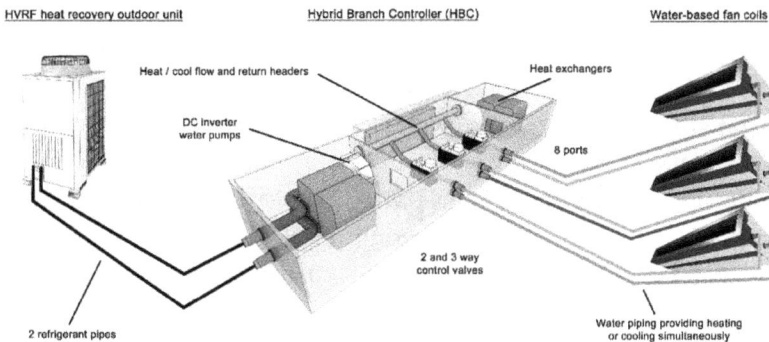

HVRF heat recovery outdoor unit
Hybrid Branch Controller (HBC)
Water-based fan coils

Heat / cool flow and return headers
Heat exchangers

DC Inverter water pumps
8 ports

2 and 3 way control valves

2 refrigerant pipes

Water piping providing heating or cooling simultaneously

Let's go deeper:

When the system kicks into heating mode, something rather remarkable happens outside. The outdoor unit starts its magic by sending refrigerant in a vapour form thanks to the compressor. This clever device takes low-pressure hot vapour – fresh from the evaporator – and transforms it into high-pressure, very hot vapour. This transformed vapour then travels through a pair of plate heat exchangers. During this journey, our refrigerant undergoes a transformation, condensing from a vapour into a liquid. Meanwhile, it generously transfers its heat to the water circuit. This hot water doesn't just stay put; it's on a mission. Propelled by pumps and guided by three-port valves, it heads straight to the indoor units. Once there, the hot water meets the fan coil unit (FCU), which spreads the heat throughout the room. It's a simple but effective way to keep things cosy. Now, let's flip the switch to cooling mode. The process mirrors heating but with a chilly twist. The vapour from the outdoor unit undergoes a transformation, becoming liquid through sub-cooling. This liquid then encounters the expansion valve, a place where it expands, becoming super cold. As it starts boiling in the evaporator, it transfers cool energy to the water and consequentially to the indoor units. But what if we need to heat and cool spaces simultaneously? It sounds like a complex job, but it's all in a day's work for our HVRF system. Here, we deploy a set of plate heat exchangers. One set takes on the heating challenge, condensing the refrigerant, while the other tackles cooling, evaporating it. This process starts always turning hot vapour into liquid, which then creates hot water for heating purposes. Following this, our liquid passes through the expansion device and then through plate heat exchangers acting as evaporators. This is where the cooling effect comes to life. The result? Cold water is dispatched to the indoor units, bringing comfort by lowering temperatures, while the low-pressure vapourised refrigerant cycles back to the outdoor unit. This marks the beginning of another round of sustainable thermal management.

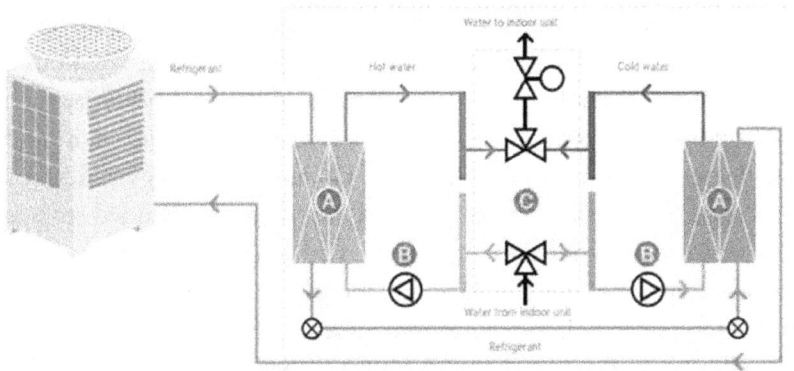

Hot water storage and its impact on reaching net zero

Different parts of the globe have different regulations and in particular the hot water storage became a "hot" matter in this sector. I attended a very interesting CIBSE webinar regarding the "heat pumps applications" and as usual the final part of it was dedicated to the Q&A section. One of the topics dedicated to the UK regulations was with regards to the hot water storage, the possibility to reduce the temperature and how CIBSE and other organizations were dealing with it[*]. These regulations can impact energy efficiency and the ability to reach the net zero targets. I believe that it's important, some time, to talk about what can be improved at regulations level and consider effective changes to help technology and engineering reach the 2035 net zero targets (in 2021, the UK government set two additional interim targets to run a net zero power system and reduce emissions by 78% by 2035). It's interesting exploring the differences between countries' regulations and their impact on reaching net zero, as well as the efforts of organizations such as CIBSE to push for regulatory change where necessary. Those differences can be attributed to various factors such

---

[*] CIBSE webinar #GrowYourKnowledge - Heat Pumps applications: beyond the catalogue data - Thu, May 4, 2023 11:00 AM - 12:00 PM BST.

as climate, plumbing systems, and cultural preferences. In some cases, hot water storage cylinders are set to a higher temperature of around 60 degrees Celsius to prevent the growth of harmful bacteria, such as Legionella. In other cases instead, a lower temperature of around 50 degrees Celsius is recommended to improve energy efficiency and reduce the risk of scalding. Reducing the recommended storage temperature will contribute to reducing the energy consumption and carbon emissions. One significant reason why a lower hot water storage temperature can help reach net zero targets is its potential to improve the efficiency of new technologies such as heat pumps; additionally, a cooler storage cylinder will lose less energy into the surrounding air as hot water stored at 50°C will have a 20% lower heat loss than a 60°C. Reducing the recommended storage temperature will make it much easier for buildings operators to switch to heat pumps as those systems work better and more efficiently when supplying lower hot water temperatures.

I know what you are thinking right now... that's great but what about the Legionnaires' risk?

Before working out how we can deal with it, just a brief reminder: lowering the recommended hot water storage temperature to 50 degrees Celsius can raise concerns about the growth of harmful bacteria, which can cause Legionnaires' disease (a type of pneumonia). Most residential properties use a significant portion of their water daily and are unlikely to have stagnant water for extended periods, such as during a typical vacation. However, commercial properties are more susceptible to risk, as they can go unused for months or experience sudden increases in water usage. This risk extends not only to the property itself but also to the associated plumbing systems. In contrast, hot water points in domestic homes are used more frequently, resulting in a lower risk. To address this issue in commercial and rental properties, the L8 Legionella regulations and the HSG274 guidelines require the properties' mangers to maintain hot water temperatures at a minimum of 60°C and "install" temperature reducers at all outlets.

In 2017, a study conducted by infection experts from Public Health England examined 99 showerheads in 82 UK properties.

They discovered that nearly one-third of them tested positive for Legionella bacteria. However, the actual number of reported cases of Legionnaires' disease was quite low, suggesting that the real danger lies in the bacteria growing to hazardous levels. The key question then becomes: How do we prevent Legionella bacteria growth?

For Legionella bacteria to reach dangerous levels, two conditions must be met:

- Stagnant water, where the water is not circulating or being replaced.
- Water temperatures within the bacteria's growth range, which is between 20°C and 45°C.

It's clear that by eliminating either of these conditions, Legionella growth can be prevented. Again, we need to remember that the temperature recommendations should have the goal of stopping the bacteria growth and not killing the Legionella bacteria, therefore reducing the storage temperature from 60 degrees to a more manageable yet still effective range can help inhibit the bacteria's proliferation without compromising the Health and Safety. As mentioned, Legionella bacteria grow between 20 and 45 degrees Celsius, which means that a storage temperature of 50 degrees Celsius can effectively prevent their growth.

Additionally:

- Regular flushing and maintenance of the hot water storage system can help to reduce the risk of Legionella growth and improve safety.
- Periodical sterilisation/pasteurisation or Anti-legionella Cycles can be performed. (Heat pump systems often have automated weekly or bi-weekly sterilisation built in.)
- Also, the new generation of thermostatic mixing valves with thermal disinfection function can further prevent the growth of harmful bacteria.

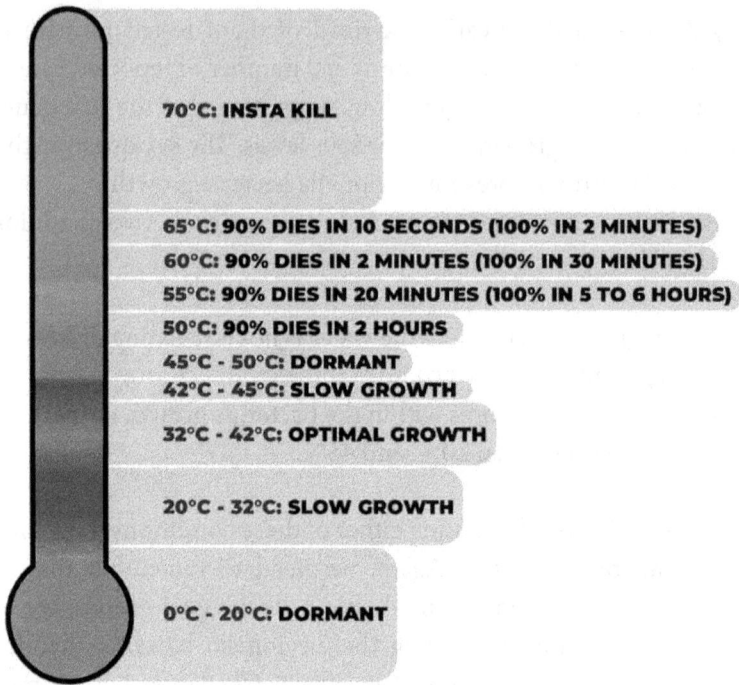

70°C: INSTA KILL

65°C: 90% DIES IN 10 SECONDS (100% IN 2 MINUTES)
60°C: 90% DIES IN 2 MINUTES (100% IN 30 MINUTES)
55°C: 90% DIES IN 20 MINUTES (100% IN 5 TO 6 HOURS)
50°C: 90% DIES IN 2 HOURS
45°C - 50°C: DORMANT
42°C - 45°C: SLOW GROWTH
32°C - 42°C: OPTIMAL GROWTH

20°C - 32°C: SLOW GROWTH

0°C - 20°C: DORMANT

The Health and Safety Executive (HSE) is responsible for setting hot water storage regulations in the UK, and several organizations are advocating for the HSE to reconsider its regulations to align with EU standards. I find interesting the idea to lower the current recommended hot water storage temperature to 52-54 degrees Celsius. As we strive to reach net zero targets, it's central to consider all possible ways for reducing carbon emissions and improving energy efficiency, promoting the use of new technologies that use different sources to produce energy compared with the conventional methods. While concerns around Legionella growth are understandable, a storage temperature of 50-55 degrees Celsius can effectively prevent its growth; therefore, adopting a lower storage temperature can potentially help to reach our net zero goals while maintaining safety standards.

What do you think? Quite an intriguing matter, isn't it?

The Rise of Heat Networks

We're quite the power users in our daily lives, aren't we? With our ever-growing energy needs, it's only logical to tap into a variety of sources. And as we become more aware of the environmental toll of fossil fuels – not to mention their looming scarcity – the quest for 'greener' energy solutions is picking up pace. Among these, district heating is emerging as a shining option.

Take hydrogen, for instance. It's like the eco-friendly cousin to gas, producing just water vapour and heat, with zero carbon emissions. Pretty neat, right? But here's the catch: most of the hydrogen we use today still comes from those same old fossil fuels. Hydrogen comes in various colours, each representing a different production method and environmental impact.

- First up, we have Green Hydrogen. This is the eco-warrior of the hydrogen world. Produced by splitting water into hydrogen and oxygen using renewable energy sources like wind or solar power, it's as clean as it gets. No carbon emissions here!
- Then there's Blue Hydrogen. Think of this as a middle ground. It's produced from natural gas, but the twist is that most of the carbon emissions are captured and stored away, rather than released into the atmosphere. It's not as perfect as green hydrogen, but it's a step in the right direction, like making a cake but swapping out some sugar for a healthier option.
- We also have Grey Hydrogen, which is the most common type currently. It's made from natural gas through a process called steam methane reforming, but without capturing the carbon emissions. So, it's like baking a cake the traditional way – it's effective, but not the best for your health, or in this case, the planet.
- And then there's Turquoise Hydrogen, a newer player on the field. It's made by breaking down methane into hydrogen and solid carbon using a process called methane pyrolysis.

Sure, hydrogen has got a bright future in our energy mix, especially for power-hungry industries, but when it comes to heating our homes, district heating might just be the smarter choice. It's all about using resources wisely and keeping that precious hydrogen for where it's most needed.

Now, let's talk heat pumps. Imagine fitting every residence with its own heat pump. Sounds good, but it's a bit like everyone making their own bread from scratch – it's doable, but not exactly efficient. This is where heat networks come into play. They're like the neighbourhood bakery of heating, offering a more economical approach through the magic of scale. Why maintain a bunch of small pumps when you can have one big, efficient one doing the job for everyone? It's cost-effective, and let's face it, a lot less hassle.

So, let's investigate heat networks, often referred to as district or communal heating. These aren't just buzzwords; they're trends shaping our industry for the years to come. Now, you might be wondering, what's the difference between district and communal heating? Well, it's pretty straightforward. Communal heating is like focusing on a small neighbourhood – it typically supplies heat to a small area, like a single building or a couple of them, often seen in multi-storey buildings or sheltered housing complexes. On the other hand, district heating is the big player. It's about spreading warmth over a larger area, connecting multiple buildings in a vast heat network. Both of these systems are what we call heat networks. And here's a fun fact: heat networks aren't a new concept. This technology has been the underdog for years, often overshadowed by more carbon-heavy local heat sources, but back in 2016, the UK Government decided to put a substantial £320 million into district heating networks over five years. In fact, these networks have become the foundation of the government's 10 Point Plan. This ambitious plan isn't just a list; it's a commitment to transform our energy sector with green technology and march towards a carbon-neutral future by 2050. So, when we talk about district heating, we're talking about a key player in our journey to decarbonize heating and revolutionize how we think about energy. Traditionally, linking up existing buildings to these networks was a tango between the supplier and customer, a purely commercial affair. New buildings, nudged by local planning requirements, were often the ones getting connected. But here's where the plot thickens: legislation and directives are setting the stage for a big shift, giving local planning teams the power to mandate connections to new heat networks. And if that wasn't enough, the government, is introducing heat network zones across the country. With Ofgem stepping in as the regulator, heat is joining the big leagues alongside electricity and gas. The stage is set, and more than 100 UK towns and cities are lined up as the first acts in this large-scale rollout.

How does district heating work?

At the core of this system is the central heat source, which can vary significantly in form, depending on the specific needs and resources of the area it serves.

The central heat source often takes shape as a large energy centre or plant room. This hub is equipped with heat generation units, which could be boilers that operate on natural gas, oil, or coal. These boilers function by heating water to produce steam or hot water, which is then distributed to various locations. A prevalent option is the use of a Combined Heat and Power (CHP) unit. There's also the innovative approach of using recycled industrial waste heat, capturing and repurposing heat from industrial processes that would otherwise be dissipated into the environment.

Once heat is generated at this central point, it needs to be distributed to the consumers – the buildings and facilities across the network. This distribution is achieved through an elaborate network of insulated pipes, specifically designed to minimize heat loss. These pipes transport the hot water or steam from the central heat source, weaving through the area much like a circulatory system. To maintain the necessary pressure and flow for the heat to reach even the most distant points of the network, pump stations are strategically placed along the route. Upon reaching a building, the heat from the network is transferred into the building's own heating and hot water systems. This critical transfer is facilitated by Heat Interface Units (HIUs). Each HIU is equipped with heat exchangers, essential components that allow for the transfer of heat from the network's hot water to the building's system, all while keeping the two water sources separate. The HIUs also include control valves and gauges, which regulate the flow and temperature, ensuring that each building receives the appropriate amount of heat. Similar to a Heat Interface Unit but with a focus on cooling rather than heating the Cooling Interface Unit (CIU) is a key component in the district cooling functions.

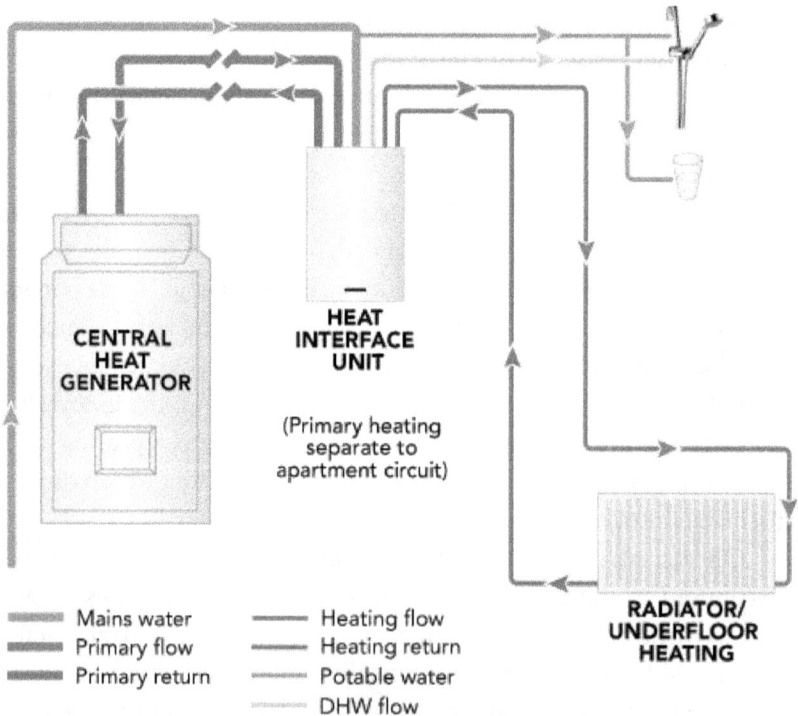

CENTRAL
HEAT
GENERATOR

HEAT
INTERFACE
UNIT

(Primary heating
separate to
apartment circuit)

RADIATOR/
UNDERFLOOR
HEATING

Mains water
Primary flow
Primary return

Heating flow
Heating return
Potable water
DHW flow

The efficiency of district heating lies not only in its centralized production but also in its capacity to utilize a variety of energy sources, including renewables and waste heat. This flexibility allows for a more sustainable approach to heating, reducing reliance on fossil fuels and decreasing overall carbon emissions. Moreover, the system's centralized nature enables more effective monitoring and maintenance, ensuring consistent and reliable heating for all connected consumers. Now that we know how a network heating works, it's easy to understand that it's not just about what's working today; it's about paving the way for the future. Imagine a world where new, innovative energy sources are being discovered and harnessed. The remarkable flexibility of district heating systems truly shines when we consider their ability to integrate with renewable alternatives. Think how the technology will improve efficiency from the geothermal energy that comes from the ground, biomass derived from organic materials, the powerful breezes of wind energy, the force of water energy, and the abundant rays of solar energy.

Sustainable heating and cooling with a district heating network

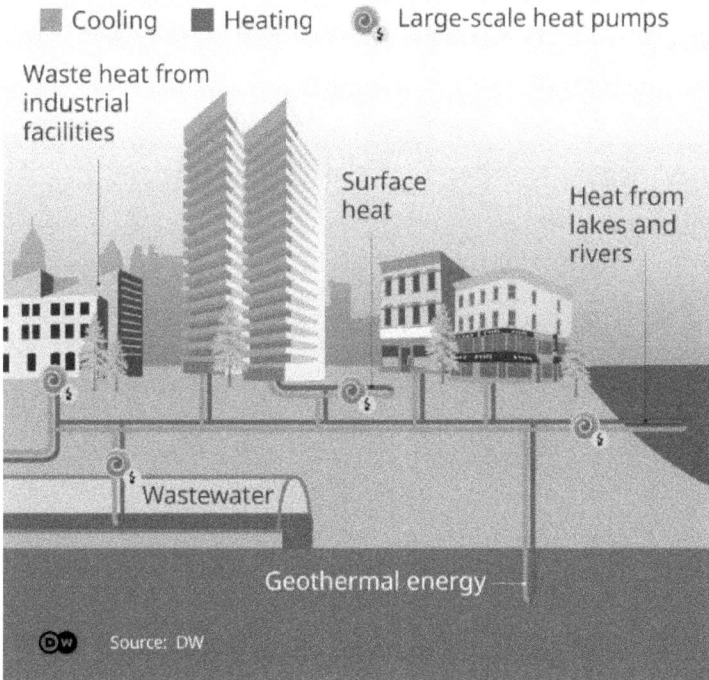

Cooling ■ Heating ◉ Large-scale heat pumps

Waste heat from industrial facilities

Surface heat

Heat from lakes and rivers

Wastewater

Geothermal energy

Source: DW

These sources are like nature's gifts – using this energy doesn't increase the concentration of greenhouse gases like CO_2, methane, nitrous oxide, and CFCs in our atmosphere. It's like cooking a meal without leaving any mess behind – clean and sustainable. When it comes to integrating these renewable energy sources with district heating, the compatibility is remarkable. Whether it's directly using geothermal heat to warm up our buildings or indirectly using electricity generated from wind and water to power heat pumps in the district heating system, the synergy is evident. So, what we're seeing here is a harmonious blend of district heating systems with renewable energy sources. This blend is not just about providing energy; it's about doing so in a way that respects and preserves our environment.

Section
6

Engaging the team in the pursuit of net zero goals

Have you heard of 'Net Zero'? Of course you have! It's like the ultimate superhero of the fight against climate change, showing us how we can all band together to make a difference!

Michael Jordan once said: "Talent wins games, but teamwork and intelligence win championships." I believe that as we continue to work toward a sustainable future, one of the most critical aspects we need to address is the involvement of our team in the pursuit of net zero goals. The role of engineers in shaping our world has never been more critical than it is today. In this chapter, I'd like to share with you some personal insights, experiences, and strategies on how we can effectively involve our engineering teams in the mission for a net-zero future.

Communicate the Vision and Set Clear Expectations

First and foremost, it's essential to communicate the vision of a net-zero future clearly and consistently. This includes setting measurable targets and defining the role each team member plays in achieving those goals. By providing a clear roadmap, you empower your team to take ownership of their work and feel responsible for contributing to the larger mission. In my experience, sharing progress updates with the team through graphs and monthly comparisons can help increase interest and passion. When team members can see the tangible improvements they are making, they become even more dedicated to the cause.

Foster a Culture of Innovation and Collaboration

We should create a culture that encourages innovation and collaboration. This means promoting open communication, providing opportunities for the team to share ideas and experiment with new technologies, and recognizing the achievements of individuals. When people feel they can contribute and make a difference, they become more engaged and committed.

I remember one project where our team was focused on understanding how mechanical equipment adjustments could impact monthly energy usage. We actively monitored the property areas' temperatures and adjusted settings where possible or implemented fan motor schedules. This collaborative and simple effort resulted in significant energy savings, reinforcing the importance of teamwork and innovation.

Invest in Education and Skills Development

Our team is the frontline for implementing solutions to increase efficiency in the buildings, therefore preparing the team with the knowledge and skills necessary, investing in continuous education and training programmes focused on sustainability and net-zero technologies will inevitably save costs and energy as well as encourage attendance at conferences, workshops, and webinars to stay up-to-date with industry trends and best practices.

Align Projects with Net Zero Goals

In order to involve our engineering teams in achieving net zero goals, it is crucial to align projects with these objectives. This means prioritizing sustainable design and construction methods, integrating renewable energy sources, and considering the environmental impacts of materials and processes.

By aligning projects with net zero goals, our engineers will naturally become more engaged in the process and motivated to find solutions that drive energy savings. This mindset will also spread among other departments and colleagues.

Celebrate Success and Share Stories

Finally, it's essential to celebrate the successes and milestones achieved along the way. Recognize the efforts and accomplishments of your engineering team, both individually and collectively. Share stories of innovation and positive change that have resulted from their work. By celebrating their achievements, you reinforce the importance of their contributions and inspire them to continue pushing the boundaries of what's possible in pursuit of a net-zero future. One memorable occasion was when our team exceeded a major energy reduction milestone and I asked to show and share those data with monthly reports so that everyone could see and acknowledge their hard work and dedication. This gesture not only strengthened team morale but also fostered a sense of pride and commitment. In conclusion, by nurturing a culture of innovation, investing in education, and aligning projects with sustainability objectives, we can inspire and motivate to drive the change we so desperately need. Let's work together to make the net-zero future a reality for all. Here's a quick tip for you, as you navigate the dynamic world of engineering, facilities management, and maintenance, the game is changing, and energy management expertise is becoming more and more important. Embrace this change and grab the opportunity to develop your skills in this vital area. Trust me, focusing on this area will not only make you a better professional but it will unlock unparalleled success in your career.

Net zero partnership

Everyone talks about it, but only few really understand what this means; however, here's the thing: we can't achieve this goal without getting seriously efficient with our energy use, and that's where consultants and end users come in. These guys are the real drivers in the net zero game, helping organizations become more sustainable and achieve their goals. With so much riding on us, it's no wonder that consultancy firms are jumping on board the net zero train. They're hustling to become net zero emissions ASAP, throwing money into making their own operations greener, and launching services to help their clients join the net zero club. Now, check this out, the EMA (Energy Managers Association) conducted a survey of around 25 consultancy service providers and client organizations to get their thoughts on how to improve the relationship between consultants and end users regarding any discipline and not only about net zero scope. The results were pretty interesting, and there are some key takeaways I want to share with you.

The question that you are probably just thinking is: How do we start and what do we need to know? Well, firstly, we need to be mindful of factors like communication and transparency when setting up these partnerships. There are also certain areas where consultants and end users can really work together to achieve great results, but, of course, there are always things to watch out for as the relationship evolves. Overall, it's clear that we need everyone on board to reach the project finish line. So, let's keep up the good work, one sustainable step at a time!

1. When choosing to engage any consultancy partnership there are several important factors to consider. In this case I want to talk you through Net Zero planning as I found it quite interesting and challenging because it's a little more detached from pure engineering or FM conventional responsibilities. Achieving Net Zero can mean different things to different organizations, and with all the nonsense out there and greenwashing, it can be difficult to know where to start. Choosing the right consultancy partner for your organization is very important and let me be clear: it is a big responsibility; in fact, it is crucial to consider their relevant experience, previous references, clear scope of

work, and clear deliverables. Service providers should offer a comprehensive set of services and expert industry knowledge, but also demonstrate humility in their approach. Net Zero requires a wide range of advisory expertise and technical support from sustainability specialists. It's important to keep an open mind when reviewing all possible solutions for Net Zero. A consultancy should be able to identify fresh ideas for consideration and show flexibility by not adhering to a standard audit process, allowing for discussions on particular industry-specific issues. Net Zero planning starts by understanding your organization's impact on the environment, agreeing on the boundaries of your climate ambitions, and devising a robust and informed strategy for action. The consultancy should be able to determine the key piece of work that your organization needs help with, such as Scope 1.2.3 emissions inventory and action plan or ESOS phases. When selecting a consultancy partner, it's important to consider where their expertise lies and to validate any claims of ability or experience. Conducting a thorough due diligence exercise can help ensure that the right people are selected for the job. Different consultancies specialize in different areas, with an engineering consultancy being better suited for delivering operational net zero from buildings, while a specialist environmental consultancy might be more suitable for calculating Scope 3 emissions. A consultancy partner should provide a comprehensive range of services relevant to the task. With Net Zero, this means covering everything from energy efficiency to renewable electricity and heat. In today's constrained market, it's crucial to have the ability to fulfil the services offered. If there is a need to make variations to the initial scope of work to better deliver the solutions, these changes should be agreed upon.

2. Where are consultancy partnerships useful?
   This is one of the questions that I always found a bit difficult
   to explain without falling into boring costs versus benefits
   discussions, but the real dilemma here is … and I will be
   completely honest, how difficult is it for a professional to
   accept the idea that in certain circumstances some help is
   needed? Someone who can help finding the best solutions
   for our organization. Yes! It's important to understand when
   it is time to seek experts in that specific sector to avoid being
   gobbled up by something which will be hard to manage,
   risking failure. Consultants offer an impartial perspective
   that can help senior management make informed decisions.
   They can identify cost and savings that may not be apparent
   to those within the organization. For example, reducing
   carbon emissions goes beyond direct operational changes. It
   includes the emissions associated with other services, such
   as those within the supply chain. For some organizations,
   Scope 3 emissions can make up the majority of their carbon
   footprint. This is especially true for companies with a large
   network of global suppliers. In these cases, seeking advice
   from consultants with supply chain expertise is crucial.
   There are some areas where things change rapidly, such
   as ESOS, Energy Audits ISO 50001, energy behaviour
   awareness, and training. There are also areas that may only
   be "visited" every few years, like EV charging infrastructure
   or renewable energy technologies such as solar panels or
   heat pumps. To tackle these areas, consultants often spend
   a significant amount of time on tasks that require in-depth
   review. As mentioned at the beginning of this chapter,
   sometimes, the required expertise is not available in-house
   and the smarter thing you can do is to choose the right
   consultancy partner that can best meet your needs.

3. Okay! We are now confident that we need partnering with
   a consultancy company. What should a good partnership
   include? Here are a few key things to look out for:

- It's important that your consultancy is open and transparent with you, sharing data and information freely and building trust through teamwork. After all, a successful project requires collaboration from both sides.
- Another key aspect of a good consultancy partnership is a commitment to embedding knowledge gained throughout the project. Your consultancy should help you learn and grow as a business, rather than just completing the one-off project.
- Finally, it's important that your consultancy is proactive in identifying and addressing any issues that arise during the project. Regular face-to-face meetings can help ensure that everyone is on the same page and working together effectively.
- Trust is vital. You'll need to build a strong relationship with your consultancy team. This trust will be crucial when it comes to accurately assessing savings, investments, and effects on reliability and productivity. To achieve this, your consultancy team will likely need to invest significant time and resources into upfront engineering studies.
- Good communication. Throughout the consultancy process, your team should be communicating regularly with you and transferring knowledge and skills whenever possible. By the end of the project, you should have a better understanding of what the consultancy team has done and be able to apply that knowledge in your business going forward.
- Overall, a successful consultancy partnership requires both parties to be committed to building trust, communicating effectively, and working together to achieve common goals. By keeping these things in mind, you can ensure that your energy efficiency project is a success and that you get the support you need to achieve your goals.

4. Something to avoid
    - Firstly, it's important to keep the consultancy team focused on where they can add the most value. For example, if you have access to internal support for administrative or data-related tasks, it's best to utilize those resources, avoiding paying for unnecessary external services. The key to this is having a clear understanding of what you need, and committing time upfront to planning how the consultancy support will work.
    - It's also important to avoid unnecessary services that could water down the quality of the service provided. Only pay for services that add real value to your operations and align with the requirements outlined in your tenders or instructions.
    - Managers may be hesitant to use consultants due to a lack of trust in the quality of work they will receive. However, even a basic understanding of the outsourced area can help you avoid false or overstated promises and prevent the waste of precious resources that could be used more effectively.
    - Finally, be mindful of vague definitions and the potential misunderstandings and assumptions the service provider can cause. Effective communication is key to ensuring that everyone is on the same page and that the project stays on track.

5. Key points
    - Understand your own skillsets and needs: Before bringing in external consultants, it's important to evaluate your own organization's capabilities and identify areas where you need additional support.
    - Consider whether better value would be gained by training up and bringing in-house. In some cases, it may be more cost-effective to invest in training and

developing your own internal resources rather than relying on external consultants. Not to mention the benefits for your personal development.

- Test a consultant's true 'expertise': Don't be afraid to ask questions and push back on recommendations that don't align with your organization's goals and values. It's important to test a consultant's expertise and ensure they are the right fit for your organization.
- Try to engage in joint planning with the consultancy team before the project begins. This can help to clarify expectations, identify potential challenges, and ensure that everyone is working towards the same objectives.
- It's important to structure the work programme in a way that maximizes the value of the consultancy team's expertise. Be clear about roles and responsibilities, and ensure that everyone is working together towards the same end goal.
- Learn the basics of the outsourced area to effectively evaluate the consultant's work and hold them accountable for delivering results.

I'm tempted to write another tome about energy, but let's be real – we want actionable insights, not a lengthy lecture. So, let me show the essence of energy strategy in today's world. There is a fine line between energy and engineering – I have noticed that many assume energy and engineering are one and the same. But here's the truth: while technical professionals and facilities managers are experts in their sector, it doesn't inherently mean they excel at energy conservation, although, an engineer with a grasp on energy concepts can be a stellar energy manager. On the flip side, an energy manager lacking this foundational knowledge is like a sailor navigating without a compass. Gone are the days when our sole concern was whether equipment worked. Today's world demands efficiency in operation, not just effectiveness. Energy strategy today is similar to the stock market – surrounded by proclaimed experts, yet only a few truly grasp its intricacies.

Whatever the organization where you work, before embarking on energy-saving missions, it's pivotal to ask:

1. How effectively are you monitoring your energy consumption?
2. Are you familiar with your energy baseline?
3. Can you pinpoint what drives energy usage in your building?

Familiarizing yourself with your building's systems is non-negotiable. This knowledge informs the interventions you choose, ensuring your efforts are targeted and effective. Begin by establishing an extensive energy tracking system. By capturing and analysing your gas, water, and electricity trends, you'll be poised to identify inefficiencies and measure the success of your conservation initiatives. Raw data is like an unsolved puzzle. Interpret it!

External temperatures often influence gas, while electricity usage might correlate with building occupancy. However, these patterns hinge on your specific setup and operations. Installing meters and sub-meters in the key areas or equipment is useful. You would link all of the meters to a central portal and the magic is done, an entire new world of information will appear where often it will surprise the user, showing the real aspect of the energy consumption.

Have a look below at some practical actions:

1. Optimize MEP systems:
   Audit and create an asset of all your HVAC equipment. Programme specific schedules to conserve energy. For instance, switch off areas where extraction systems are not necessary 24/7.
   If you have a pool, consider installing a cover. This not only reduces evaporation but also conserves heat.

Consider a gradual reduction of the pool water temperature by one or two degrees. This slight change often goes unnoticed by guests, yet can offer energy savings. Concurrently, you can reduce the ambient temperature, further adding to energy conservation.

2. Planning temperature settings for summer and winter.
3. Consider setting your heating system to a more dynamic temperature schedule, ensuring they run efficiently without compromising on heat supply.
4. Upgrade to motion sensor LED lights. This ensures illumination only when needed, cutting down on unnecessary energy use.
5. Install flow regulators for taps and showers.

The clock is ticking, and with each passing moment, the threat of irreversible climate change looms larger. If we, as a global community, are serious about averting an impending climate catastrophe, collective and accelerated action on sustainability is non-negotiable. This isn't just an abstract challenge – it's an existential one, shaping the world we'll leave behind for generations to come. At the heart of this call for action is collaboration. Alone, our efforts might seem like mere ripples, but together, we can create waves of lasting change. And it's not just about acknowledging the problem. It's about equipping ourselves with the tools and knowledge to combat it. It's an unsettling truth that human activities are the principal culprits behind our rapidly changing climate. The evidence is undeniable, and with each new report and study, the urgency escalates. The question is, are we too late? While time is of the essence, hope is not lost. This is our clarion call, not just to recognize the challenge but to rise to it. Professionals must be at the forefront of this battle. As an expert, you hold a unique and pivotal position. Bridging the gap between landlords, end-users, and supply chains, you are poised to influence and usher in a transformative shift towards a sustainable future. Every step you take, every strategy you employ, and every decision you make can set in motion a ripple effect, culminating in a healthier, greener, and more sustainable tomorrow. The responsibility is on us.

Investments for Tangible Results:

Beyond scheduled operations, consider retrofitting older systems. Newer HVAC technologies often offer greater efficiency and adaptability to varying demands.

Beyond motion-sensor LEDs, consider daylight harvesting systems and dimmable lighting solutions that adjust to natural light availability.

Beyond flow regulators, explore options like greywater recycling or rainwater harvesting to further reduce your water footprint.

Host regular workshops and bring in experts to nurture a culture of energy consciousness among staff.

Go beyond traditional methods. Explore thermal wraps, green walls, or even roofing solutions that offer superior insulation.

Annual energy audits can shed light on evolving inefficiencies. Likewise, meticulous maintenance ensures equipment longevity and peak operation.

Embracing Advanced Tech:

- Geothermal solutions
- Install demand control ventilation in the kitchen. By adjusting the ventilation based on real-time demands, you save energy and maintain optimal air quality.
- Upgrading Motor Systems. Transition from older motors to the modern synchronous reluctance magnetic system. These motors are more energy-efficient and offer improved performance.
- Room Energy Management: Consider integrating a REMS (Room Energy Management System). With this in place, equipment within guest rooms will automatically switch off or adjust settings when rooms or offices are unoccupied.

- Enhanced Electrical Efficiency: Voltage Optimization systems. By adjusting and optimizing the incoming power, these systems can instantly reduce electricity consumption and prolong the lifespan of your electrical devices.
- Renewable Solutions: Transitioning to renewables is no longer a distant dream. Solar panels, wind turbines, or even hybrid systems can pave the way to a more sustainable energy future.

Energy savings from management and efficiency programmes can be measured and verified using various methods and options, depending on the project type and available data. The primary methods include:

Key Parameter Measurement

Savings from new lighting installations are calculated by assessing changes in operating hours or power draw. The baseline consumption is determined by inputting the data and runtime of the specific equipment into a model before the project completion, typically in Excel. After implementing the new equipment, the same analysis is conducted to determine its new characteristics and performance. The normalised difference in consumption between the baseline and post-implementation constitutes your savings.

Retrofit Isolation

A meter is fitted to a circuit to compare consumption before and after an energy efficient equipment installation. An existing sub-meter or a new sub-meter measures the equipment's consumption over a set period. After the changes are made, consumption is measured again for an equivalent period. The difference in normalized consumption before and after the retrofit indicates the savings.

Whole Facility

The overall building consumption is measured before and after an energy saving installation. A submeter, either existing or new, records the property's total consumption for a designated period. After implementing the changes, consumption is measured again over a similar period. The normalised difference in consumption before and after the installation reflects the savings.

Calibrated Simulation

This method uses complex computational models to estimate the expected change in energy consumption. By analysing the consumption characteristics of the existing equipment, a model can be developed to typify consumption based on various factors. This model predicts avoided consumption and is particularly useful for scenarios involving heating, weather, and plant efficiency at different loads.

Decarbonizing and saving energy are often mentioned in the same breath, but they are not necessarily synonymous. While they are interconnected concepts, it's vital to understand that decarbonization doesn't always equate to cost savings. It's a common misconception to equate "low carbon" with "low cost". However, there's another side to this coin: if assets are not decarbonized, their future value could be compromised. For facilities managers and engineers, this discussion is crucial, especially during times of refurbishments or the installation of retro-fit systems. As professionals equipped with the right skillset, facilities managers and technical managers play a pivotal role here. We offer invaluable input into the design process since we are the ones responsible for managing and maintaining these spaces and buildings post-implementation. Our voice in this debate is not just important – it's indispensable.

In this rapidly evolving world, your energy strategy must be agile and forward-thinking. Regularly re-evaluate, refine, and reimagine your approach. Below I write what I believe is a general but effective pathway to decarbonize your building:

Decarbonization Roadmap

1. Monitoring:
   - Conduct an energy audit to understand current consumption patterns.
   - Install smart meters for real-time tracking of energy usage.
   - Utilize energy monitoring software to gather detailed insights.
2. In-House Adjustments:
   - Foster a culture that promotes energy-saving practices among all stakeholders.
   - Make necessary adjustments to equipment to enhance efficiency.
   - Optimize settings for temperature and lighting based on usage patterns.
   - Set equipment to power-saving mode during off-peak hours.
   - Implement automatic shutdown systems to minimize wasted energy.
   - Install timers or motion sensors to control lighting and reduce unnecessary usage.
   - Analyse existing systems and make improvements in alignment with the property's design and goals.
3. Developments:
   - Regularly review collected energy consumption data.
   - Set clear, achievable energy-saving targets.
   - Conduct periodic internal audits to assess the effectiveness of implemented measures.
   - Stay informed about the latest energy-efficient technologies and best practices.
   - Explore options for integrating renewable energy sources.

4. Carbon Offsetting:
   • Investigate and invest in certified carbon offset
   programmes to compensate for unavoidable emissions.

Project management

I know what some of you might be thinking. *I'm not a project manager. That's not what I do.* Well, that's kind of true. But here's the thing – every day, you're managing projects without even knowing it for both your personal life and at work. Whether they're big or small tasks, they're all projects. You have a goal, you figure out how to get there, and then you do it. One of the toughest tasks I've had was getting a hotel ready for its grand opening. Even though there were lots of experts like engineers, project managers, architects, and designers on the construction team, I was right in the middle. I had to work between the team that would run the hotel once it opened, and the team building it. My main job? To make sure everyone was on the same page and working together smoothly, while building the engineering department from scratch. Being in that middle spot can feel pretty awkward, but it's actually a great chance to learn a lot of things that can help you in your career. For example, the way engineers and interior designers work together is always something that catches my attention. A big issue they often have is about where to put things like access points for the MEP (mechanical, electrical, and plumbing) equipment. These could be in ceilings, floors, or walls. Interior designers think these access points mess up their designs and wish they could just get rid of them. But, jokes aside, there's often a bit of a struggle to find a middle ground. I believe sometimes you can find a compromise, but the main thing is, equipment needs to be reachable to work right. Putting access points for equipment in the right places is super important. It makes sure that things like electricity and water can get around the building easily. Plus, it helps when you need to fix or check on these systems later on, keeping everything safe and working well. Engineers and designers should start talking about this at early stage, ideally when they're just starting to plan the building with the architect (RIBA 2). This helps avoid problems and extra costs from trying to add access points after everything else is finished. I mentioned RIBA, but… What's that? RIBA stands for the Royal Institute of British Architects. It's a group that has a special plan for how to run building projects. This plan is like a map that shows the steps from starting a project to finishing it. Starting with RIBA's first step, which is about planning and coming up with the first ideas, is really important. It helps everyone know what they're supposed to do and how to make the building look good and work right. By following RIBA's advice from the start, the people involved in the design and construction can make sure they do a good job, talk to each other clearly, and end up with a building that does what it's supposed to do. Think of RIBA like a guide that helps everyone stay on track and avoid big problems, making the whole project go smoother. Here's a brief overview:

Stage 0: Strategic Definition – Determine the client's needs and the project's feasibility.

Stage 1: Preparation and Brief – Develop the project brief and prepare the strategy to move forward.

Stage 2: Concept Design – Develop the initial concept design to outline the project's approach.

Stage 3: Spatial Coordination – Further develop the concept into a coordinated design.

Stage 4: Technical Design – Finalize design details and specifications.

Stage 5: Manufacturing and Construction – Begin and oversee the construction process.

Stage 6: Handover and Close Out – Complete and hand over the project; address any final issues.

Stage 7: In Use – Review project performance and ensure it meets the original objectives.

Another important aspect I have learned during a construction project is how BIM became a game changer, improving the collaboration between all disciplines involved. BIM, or Building

Information Modelling, is a transformative process and technology that has revolutionized the architecture, engineering, and construction (AEC) industries. At its core, BIM is a digital representation of the physical and functional characteristics of a facility, allowing all parties to share knowledge and information about a project effectively. BIM is basically a powerful 3D software of the buildings; it's a way of getting architects, engineers, and builders to work better together, inserting all information in one place. With BIM (level 3) everyone works on the same digital model. This means fewer mistakes, improved communication among teams, and better outcomes for the entire project, as it enables everyone to visualize all the building components and their interactions. It makes it easier for everyone involved to share and update information anytime they need to. It also makes the whole building process more streamlined, cutting down on redoing work and making it easier to figure out costs and schedules ahead of time. The teams can see the building in 3D before it's even built, helping to catch any problems earlier, reducing the chances of having to make changes during construction. And when it comes to being eco-friendly, BIM is a big help too. It allows for choosing materials and designs that are better for the environment right from the start. It's a bit like a new philosophy that embodies the integration and collaboration ethos necessary for the future of construction, supporting the AEC industry moving towards more sustainable, efficient, and cost-effective project delivery. BIM is categorized into levels (maturity), ranging from 0 to 4, each representing a different degree of collaboration and digitalization:

Level 0: No collaboration. 2D CAD drafting is prevalent, with paper or electronic printouts as the primary medium.

Level 1: Partial collaboration. The use of 3D CAD for concept work and 2D for drafting and documentation. Data is shared electronically.

Level 2: Full collaboration. All parties use 3D CAD models; however, not necessarily working on a single, shared model. Information is exchanged through common file formats.

Level 3: Full integration. Architecture, engineering, and construction contribute to and access a single, shared project model in real-time, stored in a cloud-based environment. This facilitates full collaboration and is seen as the future direction of BIM.

Source: crbgroup

When approaching the practical completion date (PC) it's important to make sure that we are ready for the building handover. That's a very critical part as when the principal contractor will hand over the building to the hotel team, it means that the main construction company responsible for building or renovating the hotel has completed its work to the agreed standards and is ready to transfer control of the building to the hotel's management team. Quite scary, isn't it? Here are four main topics to keep in mind:

- Ensuring that everything has been built according to the plans, specifications, and local building codes.
- Addressing any last-minute corrections or adjustments identified during the final inspections.
- Receiving all necessary documents, such as operating manuals for equipment, warranty information, and tests certificates.
- The contractor must provide training to the hotel staff on how to operate and maintain the building's systems and equipment.

The maintenance team plays a crucial role and must have a clear understanding of their responsibilities, including how the internal environment affects the health, well-being, and efficiency. Understanding the building means knowing more than just its walls and floors. It's also about getting familiar with the complicated mix of systems that keep the building running. This includes things like the heating and cooling systems, the wires and tech that keep the lights on and computers connected, and the pipes that bring water in and out, among other stuff. Making a list of everything in the building, what condition they're in, taking advantage of open ceilings and walls to record the location and any important papers like warranties will help with keeping the building in good shape. It also helps with understanding what needs the most attention and the best way to take care of each part. Some things might need to be checked regularly, some might need a bit of work to prevent bigger problems later; in short, a clear strategy needs to be produced in advance to be prepared for the PC. Managing a project is rarely simple, and things don't always go according to plan. However, with the right attitude and motivation, it's possible to accomplish great things. And there's a huge sense of satisfaction waiting at the end of a job well done.

Experts' thoughts:

In this section I wanted to directly ask the experts about specific questions and matters that probably everyone in the sector wants to know but have never really had the chance to discuss. I have chosen various roles to understand the different perspectives and to tell us more about their personal experience.

Roxana, Assistant Chief Engineer

Ciao Roxana, could you share your journey? How did you become an Assistant Chief Engineer?

I joined the field of hospitality engineering in 2017 and started by changing shower heads and cleaning HVAC filters. I began to join experienced colleagues in their tasks, learning very quickly and gaining the tools to become a fully-fledged shift engineer. Being skilled, responsible, and conscientious, I earned the trust and respect of the Chief Engineer and moved on to the preventive maintenance programme, which I had started at my previous company. When the Chief Engineer recognized my potential, I began to spend time in the office, where I learned all about administration. At this point, I decided that I wanted to manage people and one day run my own department. I am someone who likes challenges; every time I achieved my target, I set a new one. I enjoy being in continuous mental motion, learning new things whenever I have the opportunity. I am also ambitious, always wanting more from myself. I enjoy the hospitality industry, and being a people person, I believe this has naturally led me to a management role.

In navigating a career within a field traditionally dominated by men, could you share some of the significant challenges you've encountered along your journey?

I do not see this field as being male-dominated (even though we all know it is); I know people are intrigued when they see women in the Engineering department. I believe we do not have to prove that we are good at what we are doing to anyone, only to ourselves. I have been involved in 'mending' things around the house since I was little; my father made this activity attractive. I also believe the positive experience I had at the University helped, as the teachers did not make any distinction between a man and a woman, even though we were two ladies in a class of 30 students; for them, we were all students.

Have you had any role models or mentors who have significantly influenced your career?

I believe mentorship is important for all people who pursue this career. We are not born with all the knowledge necessary to succeed in a career. My role model and mentor is the person who, without knowing me, gave me the opportunity to join this field and helped me grow, setting new challenges every day. He challenged and educated me about what engineering in hospitality means and gave me the tools to move towards management.

How do you view the impact of gender diversity in the workplace, particularly in the fields of engineering and facilities management?

I believe being a woman, and also a mother, has given me the patience to listen to all opinions in order to make the right decision. Moreover, I believe women tend to have a better attention to detail. As for gender dynamics in the workplace, especially within the engineering department, it is predominantly male-oriented primarily because there seems to be less interest among women in pursuing careers in engineering. My experiences have not only enriched my ability to weigh diverse perspectives and pay close attention to details, but also allowed me to navigate the engineering realm comfortably.

What advice would you give to young women aspiring to enter the engineering and facilities management sector?

Firstly, follow your dreams and aspirations – if you like and enjoy this field everything is possible; secondly, stick around people with experience and listen to their stories – this is the best way to capture their knowledge.

Walter, shift engineer:

Walter, what does an engineer need from their leaders that it's rare to find in the industry?

You know, it's a pretty dynamic gig here. The thing I'd really appreciate is more opportunities to learn the latest stuff. The tech keeps changing, and staying on top of it is key. But getting that kind of training isn't always easy around here. And then there's the whole budget situation. We're always trying to keep the place top-notch, right? But often, we're stuck fixing things after they break rather than getting ahead of it. A bit more cash flow to nip problems in the bud would be a game-changer. Trust is huge, too. When something goes wrong, I need to be able to jump in and handle it, no delays. It helps when the bosses let me take the reins without second-guessing every move. Communication's another thing. It's great when the higher-ups really chat with us, get our input. Sometimes it feels like they're in a different world, and that can be frustrating. Work-life balance, man, that's crucial. The hours here can get crazy, and emergencies don't exactly stick to a schedule. So, it's really cool when the bosses understand that and make sure we're not always on the clock. Safety's another biggie. We deal with all sorts of equipment and situations that can be risky. Having the right gear and a safe environment, that's super important. It's not just about getting the job done; it's about getting home safe every day. And lastly, who doesn't like a little recognition, right? Knowing that there's room to grow and that your hard work gets noticed, that's important. But not all bosses are great at showing that kind of support or laying out a path for your career. So, yeah, that's the kind of support I'd really like to see more of around here.

What would you change if you were in charge?

If I were running the show here, the first thing I'd change is how we handle fixing stuff. I'd push for more of a 'fix it before it breaks' approach. It's way better to catch a small issue before it turns into a big headache for everyone, especially our guests. I'd also splash out on some new tech and tools. You know, the latest and greatest. It makes our job easier and it's usually better for the environment too. Plus, the guests love it when everything's state-of-the-art. Training is another big one. I'd make sure everyone on my team is up to speed with the latest in maintenance and repair. It's not just about fixing things; it's about understanding how everything works, so we can solve problems faster and smarter. Then there's the team spirit thing. I'd work on building a really solid crew. You know, where everyone's got each other's back and we all chip in to help out. Makes a huge difference in how the day goes. Safety first, always. I'd double down on making sure we've got all the right safety gear and procedures in place. No cutting corners when it comes to keeping the team safe. And lastly, I'd open up the lines of communication. Not just with my team, but with the other departments and the big bosses too. More talking, more listening, and a lot more understanding of what's going on around the place.

What are the main three things that motivate you every morning?

So, what gets me up every morning in this job? Well, three things really stand out: First, it's the surprise factor. No two days are the same here. One day, it's a leaky tap in room 304, the next, it's a full-blown power issue in the dining hall. Keeps me on my toes, and I love that. It's never boring, and I get a real kick out of solving these problems. Makes me feel like a bit of a hero, you know? Then, there's the crew. We're like a little family. We've seen it all, and we've got each other's backs. There's something about working together, sorting out issues, sharing a few jokes along the way. It's not just a job when you're working with friends. Makes the day fly by, and honestly, it's a lot of fun. And lastly, the guests. You see them come in, tired from their travels, and you know your work is making their stay comfortable. When everything's working smoothly, they can relax and enjoy their time. Sometimes, you get a thank you from them, and that feels pretty great. It's about making their experience a little bit better, and that's a big deal for me.

What stops you from progressing in your career?

So, about climbing that career ladder. You know, when I think about it, it's a bit of both – my choices and other factors. Firstly, there's the whole 'comfort zone' thing. I won't lie, I like where I am. I know the ins and outs of this place, and there's something nice about that familiarity. Sometimes I think, "Should I go for that higher-up position?" But then, the thought of leaving my crew, dealing with more admin stuff, less hands-on work – it doesn't always appeal to me. Then, there are those external roadblocks. Like I mentioned, getting more advanced training isn't easy. It takes time and money, and sometimes the hotel's just not in a position to support that. And the opportunities for moving up – they're not always there. In a big hotel, those upper positions don't open up often, and when they do, it's competitive. Time's another big factor. This job's demanding. Some days I'm putting out fires (not literally, thankfully!) from the moment I clock in. When you're that busy, finding the time and energy to focus on your career growth can be tough. So, is it a choice? Kind of, yeah. I like what I do, and sometimes staying put feels right. But there are those moments when I wonder what if, you know? But then there are those external factors that make moving up a bit more complicated than just wanting it.

Marco, Chief Engineer:

Could you share your career journey leading up to becoming a Chief Engineer in a hotel?

Although my experience in the maintenance/construction field started in 2013 on office buildings and though this set the base of my knowledge on maintenance, my career in the field truly started in 2018 when I first joined the maintenance in a hotel. I must admit that luck played a part in boosting my career as there was an opening for Assistant Chief Engineer after just eight months of me joining the team. Although it was a challenge stepping up into a position where most of the team had more seniority over me, this was one of the keys in my development by placing myself in a position where my leadership skills had to be strengthened to overcome the fact that the most "junior" engineer was now "in charge", my time as Assistant Chief Engineer is where I started being involved with the managerial side of engineering, this quickly awakened my interest in the field. Self-awareness and understanding my own skills played a big part in my next step, the luxury hotels journey begins. Rather than looking for the next title, I chose to remain on the same level in order to divert my career into the luxury hotels market and therefore adapt to the changes of this new adventure before taking on extra responsibility on a new unknown.

At this stage is where I felt my curiosity in the field coming to a new level and the focus of my career diverted from just dedication to my work to self-development, while feeding this need of understanding new systems and how wide the field of engineering really is. Obviously, it goes without mention that this journey would not be possible without strong leadership skills, and I believe the fact that my leadership skills have been built in different fields of work has helped me develop myself not only by understanding how to lead my department, but also, understanding how other departments run and work. In August of 2023, after five years of dedication, research and "getting my hands dirty" the big moment arrived and I received the title of Chief Engineer; this journey is most certainly far from over. However, so far, it has been a great journey full of great people and great leaders along the way that helped me arrive where I am today. I have mentioned before luck playing its part in boosting my career, this goes mainly towards the great leaders I had along the way, from mentoring to letting me find out the extent of my own skills.

What aspects of your daily work present the most challenges?

For myself the biggest challenge presents itself on the urgency behind the resolution of works, a hotel's success is only as good as its guests' enjoyment and satisfaction. This means that troubleshooting needs to be quick, smooth, and efficient with the priority in mind being the guest. Now, different people will understand the word challenge differently, with some seeing it as a bad thing, where others will see this as a good thing. Personally, I see this word as a good thing and this specific challenge has helped throughout my career by allowing me to think outside the box, produce new methods and ideas on how to approach different scenarios and train my teams accordingly.

What aspects of your job do you find the most frustrating?

Most definitely the finance aspect. Leaving aside budgeting and forecasting, which in itself can be frustrating, this is a department that does not directly produce a profit, and at the same time a department that deals with urgent matters that can be quite costly while needing urgent attention; the bottom line is for a business to run it needs a profit. This area creates the most frustration to me; however, I have found that building a good case with great detail and raising awareness as to possible outcomes and reasoning for expenditure definitely will aid a quicker approval for such costly projects.

In your opinion, what are the three magic skills that a successful Chief Engineer should possess or develop?

Although it is difficult to reduce this to three key skills, I believe Dedication, Resilience and Organisation, these three aspects will give you a base and lead to other skills required both in terms of developing yourself as well as performing in the role. I would put emphasis on organisation as the pillar to build other skills – to be a successful

Chief Engineer you need to continuously learn and be open to new products and ideas and in the era of innovation we currently live in, organising your time not only to perform your daily task and projects, but it is also KEY in having the time to investigate and learn what the future can bring to the industry as well as build your own knowledge.

How do you balance the technical and managerial aspects of your role as a Chief Engineer in a hotel setting?

At the risk of repeating myself, I believe organisation is the key to both these aspects, having strong procedures and schedules in place and setting time to ensure these are being followed, as well as ensuring you have time aside to develop your technical knowledge. On the past two years I have found a process that works for me to build my technical knowledge; this consists of taking notes during the week on areas of improvement and set time aside to focus on those areas. Obviously, there are times where I have time set aside and no notes; however, these are the times to dive into new systems and processes in the industry to expand my knowledge beyond my day-to-day role and understand the impact and feasibility this would have on the industry.

Sylvie, General Manager:

How significant is the impact of the engineering department for the business?

The engineering department is absolutely crucial for the success of our hotel. It all starts with Maintenance and Facility Management, ensuring everything operates flawlessly and maintains our high standards. This directly influences Guest Satisfaction, as guests expect a seamless and comfortable experience. Moreover, our focus on Energy Efficiency and Sustainability is not just about reducing costs; it's also about meeting the expectations of environmentally conscious guests. We're committed to balancing eco-friendliness with economic efficiency. Safety and Compliance are other key areas where our engineering team makes a significant impact. They ensure that our hotel adheres to all safety regulations, providing a secure environment for both guests and staff. Lastly, when it comes to Capital Expenditure and Project Management, their expertise is invaluable. They lead the way in managing renovations, new installations, and technology upgrades, ensuring these projects are completed on time and within budget. In essence, the engineering department is the backbone of our hotel operations, integrating technical expertise with guest-centric services.

What indicators do you follow, to determine the strength of your engineering team?

To gauge the strength and effectiveness of our engineering team, we closely monitor several key performance indicators. The 'Voice of Guests' stands out as a primary metric. We attentively listen to our guests' feedback, both through direct comments and structured surveys. Their experiences and satisfaction levels provide invaluable insights into how well our team is performing. Additionally, conducting daily room checks is a routine yet crucial part of our assessment. It allows us to maintain our high standards of quality and functionality in all guest areas. Audit scores also play a significant role. These scores give us an objective measure of our compliance with industry standards and best practices. Another vital indicator is the loyalty of our customers. Repeat visits and sustained customer relationships are strong testaments to the effectiveness of our engineering team in creating a welcoming and well-maintained environment. Lastly, we keep a close eye on the spend variable in maintenance and the frequency of call-outs. These financial metrics help us understand the efficiency and cost-effectiveness of our team, ensuring we maintain a balance between quality service and fiscal responsibility. All these factors combined give us a comprehensive view of our engineering team's strength and areas for improvement.

There's a perception that engineering in hospitality is often overlooked. Why do you think this happens, and do you agree with this viewpoint?

To me, Angelo, 80% depends on the General Manager's approach, but there are also other factors at play. A significant reason is the behind-the-scenes nature of engineering departments. They work diligently to ensure the smooth operation of the hotel's infrastructure and systems, but their efforts aren't as visible to guests as those of front-of-house departments like reception or food and beverage. Consequently, their critical contributions might go unnoticed, leading to a perception of lesser significance. Another factor is the limited guest interaction of the engineering team. Unlike departments such as guest services or food and beverage, our engineering staff primarily focuses on maintaining and managing the property's physical aspects. Their work is essential for creating a comfortable and safe environment, but it doesn't directly engage guests, which might affect their visibility and perceived importance. There's also a cost-centric perception associated with engineering departments. They are often seen as focused on cost control and maintenance, which can lead to a misconception that their role is more about reducing expenses than enhancing guest experiences. However, it's crucial to understand that while maintaining cost efficiency is important, the engineering department's contribution to guest satisfaction, safety, and sustainability is equally vital. Lastly, the lack of awareness among some guests and even hotel management personnel about the scope and complexity of the engineering department's responsibilities contributes to this perception. A more comprehensive understanding of their multifaceted role would help in appreciating their importance in the hospitality industry.

The role of the maintenance team is evolving beyond just replacing some bulbs. How do you make sure of the expertise required when hiring for these positions?

The role of our maintenance team has indeed evolved significantly, going far beyond basic tasks like replacing bulbs. To ensure we hire individuals with the necessary expertise, we place immense importance on the first three days of their journey with us. This period is dedicated to a comprehensive induction process. We introduce the new team members to their specific job roles, familiarize them with the environment they'll be working in, and provide knowledge about our sister hotels. This broader understanding is crucial for them to grasp the standards and practices across our properties. Understanding their job tasks from the get-go is vital. We believe in setting clear expectations and providing the necessary tools and information to meet those expectations. One of the key elements in our hiring process is assigning a mentor for the first two months. Yes, there is a cost associated with this mentoring approach, but we see it as an investment. The long-term return on investment is significant – not just in terms of maintaining high standards in maintenance but also in building a team that's skilled, confident, and capable of handling the complex demands of modern hospitality maintenance. This comprehensive approach ensures that our maintenance team is well-equipped to handle their evolving roles effectively.

How do you envision the future responsibilities and skills required for engineering staff in the hospitality industry?

Looking towards the future, I see a dynamic evolution in the roles and skills required for our engineering staff in the hospitality industry. A major focus will be on Sustainable Technology and Energy Management. As we move towards a more environmentally conscious world, it's imperative that our team is skilled in implementing and managing sustainable practices and energy-efficient technologies. Digitalization is another key area. The rapid advancement of technology means our engineering team must be adept at integrating and managing digital systems, from guest services to operational controls. Cybersecurity and Data Protection have become increasingly crucial. As we handle more guest data and digital processes, the responsibility of protecting this information from cyber threats falls significantly on our engineering team. Integrated Facility Management will also take centre stage. The ability to manage and optimize various systems and facilities in a cohesive manner is essential for operational efficiency and guest satisfaction. Soft Skills and Customer Service are going to be a MUST, especially in the post-Brexit landscape. The engineering team's ability to interact positively with guests and provide excellent service will be just as important as their technical skills. Finally, Adaptability and Lifelong Learning are key. The industry is constantly changing, and our team must be able to adapt and continuously update their skills to stay ahead of these changes. This approach to ongoing learning and adaptability is what will truly set our engineering staff apart in the future.

Andrea, Mechanical Design Engineer

Andrea, how are the building engineering projects changing? Compared with a few years ago?

In recent years, building engineering projects have undergone significant transformations, largely influenced by technological advancements, sustainability concerns, and shifting user expectations. One notable change is the increased integration of smart building technologies, optimizing energy usage and enhancing occupant comfort while enabling remote monitoring and control. Since the onset of the COVID-19 pandemic, there's been a heightened interest in indoor air quality among building end-users. This has spurred the implementation of advanced ventilation and filtration systems across various sectors, from educational buildings to residential properties and office/commercial spaces. Additionally, the pandemic has catalysed the adoption of flexible and adaptable building designs, accommodating specific functionalities like dedicated office/study spaces and gym rooms within residential properties. Sustainability has also become a central focus in building engineering projects, with a particular emphasis on reducing carbon emissions associated with heating and hot water systems. Over the past few years, there has been a notable increase in interest from customers regarding renewable energy options for heating and hot water. The market for renewables has expanded significantly, with heat pumps now capable of addressing challenges like higher flow temperatures. However, one ongoing challenge for design engineers is the cost disparity between gas-based systems and electricity/renewable systems, particularly notable in existing buildings where the upfront investment for transitioning to renewable energy solutions may not always be economically viable compared to the lower initial costs associated with gas systems. Furthermore, consultants are increasingly required to account for the environmental impact of their projects, leading to the adoption of more sustainable and efficient design practices. For instance, replacing traditional chemical and corrosion control products with innovative side stream filtration not only conserves water but also improves equipment lifespan and reduces chemical waste. Overall, there's a growing awareness among end-users, occupants, and the construction industry as a whole regarding the importance of safe, environmentally sustainable buildings with minimal energy consumption. This shift in mindset is driving the adoption of innovative technologies and design strategies aimed at creating healthier, more efficient built environments.

How do these changes affect costs and expectations from clients?
The incorporation of smart building technologies, advanced ventilation, and filtration systems, as well as renewable energy solutions, often entail higher upfront costs compared to traditional building systems. While these investments may result in long-term operational savings and environmental benefits, they can initially impact project budgets and timelines. Therefore, clients may need to adjust their financial expectations and timelines to accommodate these upfront investments in exchange for long-term benefits. Moreover, the increased emphasis on sustainability and occupant health and well-being may also lead to higher client expectations regarding the performance and efficiency of building systems. Clients are increasingly seeking energy-efficient, environmentally sustainable solutions that prioritize occupant comfort and safety. This shift in expectations necessitates close collaboration between clients, engineers, and other stakeholders

to ensure that project goals align with budgetary constraints and performance requirements. While renewable energy options may offer long-term cost savings and environmental benefits, the higher upfront costs associated with these systems may deter some clients, particularly in cases where the financial return on investment is not immediately apparent. Consequently, clients may require thorough cost-benefit analyses and clear explanations of the long-term value proposition of sustainable building solutions. To conclude, the changing landscape of building engineering projects influences costs and client expectations by necessitating investments in advanced technologies, sustainability measures, and occupant-centric design principles. Clients must balance short-term budget considerations with long-term sustainability goals, requiring collaboration with experienced engineers to achieve cost-effective, high-performance building solutions that meet their evolving needs and expectations.

How critical is efficiency nowadays?

Efficiency in building engineering projects is more critical than ever before due to several factors driving the industry's evolution. Firstly, the focus on sustainability and environmental impact necessitates the optimization of building systems to minimize resource consumption and carbon emissions. Energy efficiency plays a crucial role in reducing a building's carbon footprint, mitigating its contribution to climate change, and meeting increasingly stringent regulatory requirements and sustainability standards. Secondly, the rising cost of energy underscores the importance of efficiency in building operations. Energy-efficient systems not only reduce utility bills but also enhance overall operational cost-effectiveness, providing long-term savings for building owners and occupants. Moreover, the demand for occupant comfort and productivity underscores the importance of efficient building design and operation. Well-designed HVAC systems, lighting solutions, and building envelopes contribute to improved indoor environmental quality, enhancing occupant satisfaction, health, and well-being. Furthermore, efficiency is critical for ensuring the resilience and adaptability of buildings in the face of evolving climate conditions and societal needs. Efficient building systems are more flexible and responsive to changing environmental conditions, enabling buildings to maintain optimal performance and occupant comfort under varying circumstances. In summary, efficiency is paramount in modern building engineering projects due to its impact on sustainability, cost-effectiveness, occupant satisfaction, and resilience. As the industry continues to evolve, maximizing efficiency in building design, construction, and operation will remain a top priority for engineers, architects, and building owners alike.

What should an exemplary new building have? And what certifications?

A perfect new building should embody several key characteristics to ensure it meets the highest standards of sustainability, efficiency, comfort, and functionality. The building should be designed to minimize energy consumption through efficient HVAC systems, high-performance insulation, energy-efficient lighting, and appliances. Passive design strategies, such as orientation, shading, and natural ventilation, should also be incorporated to reduce reliance on mechanical systems. Indoor air quality is critical for occupant health and comfort. The building should have adequate ventilation systems, air filtration, and pollutant control measures to ensure high indoor air quality. Sustainable building practices should be integrated into every aspect of the building's design, construction, and operation. This includes using environmentally friendly materials, optimizing resource use, minimizing waste generation, and incorporating renewable energy sources such as solar panels or geothermal heating. Also, implementing water-efficient fixtures, rainwater harvesting systems, and greywater recycling can significantly reduce water consumption and promote sustainable water management practices. The building should be accessible to people of all abilities, with features such as ramps, elevators, wide doorways, and accessible restrooms. Accessibility standards should be met or exceeded to ensure inclusivity and accommodate diverse needs. The building should be resilient to climate change impacts, natural disasters, and other potential hazards. This may include features such as robust structural design, flood-resistant construction, and backup power systems to ensure continuity of operations during emergencies. The building should be designed to accommodate evolving needs and uses over time. Flexible floor plans, modular construction techniques, and adaptable infrastructure systems can facilitate future renovations and repurposing without significant disruption or expense. The building should prioritize occupant comfort and well-being by providing optimal thermal comfort, daylighting, acoustics, and ergonomic design features. Spaces for relaxation, collaboration, and social interaction should be incorporated to promote a sense of community and wellness. In terms of certifications, several internationally recognized green building certifications can validate a building's sustainability and performance. Some of the most common certifications include:

LEED (Leadership in Energy and Environmental Design): Developed by the US Green Building Council, it is now a great validation method in the UK too. LEED provides a framework for designing, constructing, and operating green buildings across the globe.

BREEAM (Building Research Establishment Environmental Assessment Method): BREEAM is a widely used green building certification system developed in the UK, focusing on sustainability and environmental performance.

WELL Building Standard: WELL focuses on promoting occupant health and well-being by addressing factors such as air quality, water quality, lighting, comfort, and mental well-being.

Passive House: Passive House certification focuses on achieving ultra-low energy consumption and high levels of comfort through rigorous design and construction standards.

These certifications provide independent verification of a building's sustainability and performance, helping to demonstrate its commitment to environmental responsibility and occupant well-being.

James, Regional Director of Engineering

Could you share your career journey leading up to becoming a Regional DoE in a hotel?

I like to fix things, it's in my nature; when I was young, I was always trying to work out how things worked, so tended to take them apart, only to find they didn't always go back together so easily. Here is where my journey started! When I left school, I had an offer of joining a training scheme that could lead to an apprenticeship. This led me to start working in hotels by chance, securing one of two training scheme placements in the engineering department in a local hotel and a well-known hotel chain. When I completed the training scheme placement, I was offered an apprenticeship in plumbing by the hotel, of which I went on to successfully complete. As I served my time as an apprentice and beyond, I found myself working through the various roles in the engineering department, roles ranging from room technician, shift engineer, plumber, pool plant technician and eventually plant engineer. As plant engineer, this is where I learnt the most about the hotel facilities, and how important it was to look after the key assets to ensure the hotel could operate 24/7/365. I was also often found working on special projects in the hotel, including painting and decorating or fit outs of the back of house offices when any changes were needed, anything really that needed doing, I would be at the front of the queue wanting to be involved. I believe it was this kind of attitude that helped my career grow within the function. I was sent on numerous training courses throughout my time, ranging from many other technical courses to grow my skills and technical knowledge along with training and development in people development skills and business acumen that would all lead me to become part of the Engineering Function and key leadership team.

What aspects of your daily work present the most challenges?

Finding ways to communicate often complex and technical issues to key stakeholders so that effective and timely decisions can be made when looking to ensure all stakeholders can achieve the desired outcome. I also see the current knowledge and skills gap in upcoming or potential new leaders of the function; to this end I have found others looking to me as the subject matter expert when I don't see this in myself. As a result, I have found myself becoming a mentor, either indirectly or on referral or request. This has become as rewarding as doing the job itself at times.

What aspects of your job do you find the most frustrating?

This is a difficult question to answer, I have found as I have developed over my career, that most things no longer frustrate me, I now see most things as challenges to overcome.

In your opinion, what are the three key skills that a successful Regional DoE should possess or develop?

1. Communication – Having the ability to explain findings and solutions of often complex and technical issues to different audiences so that they understand the cause and effect and the benefit to the business by addressing what you are raising.
2. Problem solving – An aptitude and resilience to problem solving and a desire to maintain an up-to-date technical knowledge of often complex FM services and MEP.
3. Interpersonal and people skills – Developing your people and social skills or intelligence. Reading the signals that others send and interpreting them accurately to form effective responses.

How do you balance the technical and managerial aspects of your role?

By using the skills I have developed over many years to ensure I have prioritised the workload to meet the needs of the business, not just in the short term but in the medium to long term. This is crucial as guardian of an asset to ensure challenges are proactively managed out of the FM environment with the goal of protecting an asset, owner investment and ensure a safe and functional environment of our customers. This is to ensure key stakeholders are engaged at the right time to enable them to make informed decisions about any challenges faced, so as to ensure appropriate resources are made available to support the growth of the business.

Renata, Sales Manager & Sustainability Champion

Renata, in your role as a sales manager, how do you integrate sustainability, especially when addressing clients who might prioritize luxury over eco-friendliness?

Sustainability is part of all my sales pitches, presentations and proposals, but the depth of the topic always depends on my client's values and interest. Nowadays, most corporate clients have their own ESG goals with a major focus on travel, hence sustainability tends to be one of the most important parts of my presentations. Even though I cannot say that sustainability is as important as price when it comes to decision making, a certain degree of eco-friendliness is required to win any corporate business. When it comes to individual travellers, the picture is less optimistic, especially for those travelling for leisure. In our experience, holiday makers are more likely to prioritize comfort over eco-friendliness. This is understandable, considering that holidays are rather rare and expensive for most, hence these guests want to have the best possible experience without any compromise. A great example for this is our key card-controlled A/C. In an ideal world, our guests would always take their key cards with them when they leave the room, therefore the A/C shouldn't run while there is no one in the room. However, our guests are very creative and replace the key cards with other plastic cards (e.g. credit cards) to ensure that the temperature will be at their desired level as soon as they come back to their rooms. Unfortunately, this means that the A/C is running throughout the entire day in an empty room.

In your experience, how has the emphasis on sustainability impacted the hotel's brand and appeal in the market? Do you find that it's an effective selling point in today's hospitality industry?

Absolutely! In fact, it's becoming one of my most effective selling points. There are hundreds of 4- and 5-star hotels in London with good location, spacious rooms, and great amenities, but there are not many hotels with impactful sustainability action plans. Our industry doesn't have any significant sustainability-related legislation yet, and we don't have a widely recognized accreditation body, hence there is absolutely no consistency amongst the hotels' sustainability actions. To put it into perspective, some hotels advertise themselves as an eco-friendly establishment, because they have a bug hotel, which is a man-made structure created to provide shelter for insects, but they still have single use plastic or don't have LED lights due to the cost. Don't get me wrong, I have nothing against bugs, but that's not where a sustainability journey should end. Having a robust, meaningful ESG plan with clear goals and results helps us make the difference as a hotel and helps me as a salesperson to differentiate our hotel from the competitors.

In your opinion, how does the engineering and maintenance aspect of the hotel contribute to making your job easier?

I always think about the hotel as a machine – if one part is broken, the whole machine stops working. So, in my perspective, all departments are equally important to keep the hotel running on a daily basis. However, when it comes to long-term goals, engineering plays an extremely important role. There are new hotels opening every day with brand new facilities and the newest technologies. If we want to stay competitive, we need to focus on maintenance and continuous revival, otherwise we will lose our guests to the shiny new hotels where everything still works perfectly.

As a sustainability champion, what are the biggest challenges you face in promoting sustainable practices within a luxury hotel environment, and how do you address these challenges?

My biggest challenge is finding a discreet yet effective way of educating our guests. I don't want to make them feel that we give them too many instructions or we are constantly watching over their shoulders, but we need our guests to help us achieve our ambitious sustainability goals. For example, we now have fabric laundry bags that are probably too good looking, since we lose hundreds of them every month. While our fabric laundry bags are reusable and much more sustainable than the plastic ones, the production of the fabric bags has a much larger carbon footprint. It sounds trivial, but how do we ask our guests nicely not to take our laundry bags home?!...

Could you share some innovative ways or technologies that your hotel has adopted to enhance sustainability without compromising the luxury experience for your guests?

There are plenty of invisible innovations in our hotel. We have LED lights, CHP, ARC system for chiller optimization, energy saving system in our plant room, pool plant room inverter, water pressure controller in our taps and showers, dual toilet flush – just to name a few. These are all very impactful initiatives that our guests will never notice since they are not affecting their experience in any way.

Gregor, Chief Technical Officer

Could you share your career journey leading up to becoming a CTO?

Following graduation from Glasgow Caledonian University with a Masters Degree in Construction Project Management, I started my career in Hong Kong as a graduate quantity survey. I didn't stay in Hong Kong very long as I quickly realised you needed to earn a lot of money in that City to have a comfortable standard of life and that was years away for me. When I returned back to Glasgow, I applied for multiple jobs in all the medium to large construction companies across the Glasgow and Edinburgh central belt and didn't get one response or interview. I guess you could say there was a downturn in the industry at that time so less jobs available for inexperienced graduates or my CV wasn't very good! While going through the job-hunting process, a recruitment consultant called me out of the blue one day and asked if I wanted to go to this interview with a hotel company and whether I would be interested working for a client. At that time, I had no idea what I was doing and just wanted to get a job so went for the interview as an assistant project manager. Shortly after joining this company, which was a relatively new company in 2003 with approx. seven new build hotels (all built within the previous three years), the owners realised there was a number of property and FM related actions that needed someone to oversee so my role quickly changed to property management as we added more hotels to the portfolio. In 2011 I decided for a change and I had progressed to the role of Group Property Manager where I was responsible for all FM, statutory compliance, capex, and Health and Safety. Following three years of exposure and fantastic experience, my previous organization successfully won a large contract of 22 Holiday Inn and Crowne Plaza Hotels and in 2015 they invited me back to head up all capital works and form our Capital team focusing on delivery of project management or refurb, restoration and FF&E procurement projects. I held that position till 2018 before stepping up into my current role picking up responsibility of the Capital, FM and H&S departments. I also chair our Environmental Committee which forms part of our ESG strategy.

What aspects of your daily work present the most challenges?

Because we are a management company with approx. 25 different owners, we need to ensure that the delivery of services and our performance is consistent and of a high standard or it damages our reputation and potential future management contracts, and that responsibility sits with me.

What aspects of your job do you find the most frustrating?

Being let down by consultants and contractors. If you are paying for a service from a specialist consultant and they fail to deliver, this delays projects, adds cost, and because we are typically managing projects on behalf of multiple different owner groups, it may damage my team and my own reputation. Choice of contractor is critical and in recent years the quality of workmanship from many we have dealt with has been poor and because of this we have unfortunately parted company with several contractors.

In your opinion, what are the three key skills that a successful CTO should possess or develop?

Given the number of balls being juggled and plates typically spinning, I think the most important skill is to be organised and well-structured. The second key skill is to be reliable and efficient. We answer to multiple stakeholders including the owners of the hotels we manage, the hotel GMs and hotel teams that rely on our technical knowledge and skill, the rest of the management company, and our guests, and the assistance we provide must be consistent and of a good quality standard that all parties can rely upon. Another vital skill is to be dynamic. Every day is different to the next and while the list of priorities can often shift on a daily basis depending on the needs of our owners, hotels and our guests.

How do you balance the technical and managerial aspects of your role as a CTO in hotels setting?

I rely on key people within my team and also consultants we partner with to assist with technical aspects, but because of my background I still find myself engrossing myself with technical detail. This probably isn't the right managerial approach as you should trust the info you receive from others and rely upon it, but I guess I will always have a thirst to learn more. I still very much enjoy rolling my sleeves up and getting involved at site level and assisting and supporting my team wherever possible.

Ingrid, General Manager

How significant do you think the impact of the engineering department is on your business?

I believe the engineering department is fundamental to the success of a hotel, serving as the backbone that ensures the integrity and functionality of the entire establishment. It provides the essential structural foundation critical for the smooth operation of the hotel, from maintaining the physical infrastructure to ensuring the efficient performance of all systems within the building. This includes the upkeep of electrical systems, plumbing, heating, ventilation, and air conditioning, all of which are vital for creating a comfortable and safe environment for guests. Beyond mere maintenance, the engineering department also innovates in energy management and sustainability practices, further underscoring its importance. Their work directly impacts the guest experience, influencing everything from the reliability of services to the overall ambience of the hotel. Without the expertise and dedication of the engineering team, a hotel cannot hope to meet the high standards of operation and guest satisfaction that are key to its success.

What indicators do you look for to determine the strength of your engineering or maintenance team?

The engineering department is critical to a hotel's success, underpinning its operations with essential maintenance and repair work. Key indicators of a strong engineering department include quick response times to issues, high completion rates of both planned and reactive maintenance tasks, minimal equipment downtime, maintenance of an accurate asset inventory, proficiency in daily issue resolution, and the ability to tackle complex faults effectively. These metrics collectively gauge the department's efficiency, responsiveness, and impact on the hotel's operational integrity and guest satisfaction.

There's a perception that engineering in hospitality is often overlooked. Why do you think this happens, and do you agree with this viewpoint?

I personally think that the engineering department is indeed overlooked, due to the lack of understanding of its daily tasks which do not consist of the mere changing of light bulbs. When the department's responsibilities and contributions are not fully understood, its critical role in maintaining operations and infrastructure may be underappreciated.

Are you aware of the level of expertise required when hiring for these positions?

Nowadays, the criteria for hiring for these positions extend far beyond basic technical skills. A comprehensive and diverse skillset is now a prerequisite, encompassing not only technical proficiency but also advanced problem-solving capabilities, a thorough understanding of the latest equipment and technologies, and a robust knowledge base in specialized areas. This includes expertise in preventive maintenance, which is crucial for foreseeing potential issues and mitigating them before they escalate. Diagnostic skills are also paramount, enabling the identification and resolution of complex problems efficiently. Additionally, a deep understanding of sustainability practices is increasingly important, reflecting a shift towards more environmentally responsible operations. Familiarity with IT systems, given their integral role in modern infrastructure management, is another critical component. Moreover, knowledge of Health and Safety standards, Fire Safety protocols, and compliance with regulations ensures that operations not only run smoothly but also adhere to legal and ethical standards. The demand for such a comprehensive skillset underscores the evolving nature of these roles and the importance of a multifaceted approach to hiring in order to meet the complex demands of today's operational environments.

Considering the evolving role of maintenance teams, how do you predict the future responsibilities and skills required for engineering staff in the hospitality industry?

In the future, engineers working in the hospitality industry, may need to adapt to increasingly sophisticated technologies. Responsibilities could include managing smart building systems, implementing energy-efficient solutions, integrating automation for improved guest experience. Skills such as data analysis, cybersecurity and a strong understanding of sustainable practice may become more crucial as the industry continues to advance.

Conclusion

I really hope this book or manual has given you a clearer picture of what you need to focus on to do well in whatever area you're passionate about, and if not... I'm sorry, I gave my best!

At a symposium I attended not too long ago, something struck a chord with me. It was about how businesses are starting to really see how much difference it makes to have people who know their stuff technically. Artificial Intelligence or big data analysis can undoubtedly assist, but achieving goals becomes challenging without sufficient knowledge. Leaders are waking up to the fact that we need more training, and it's super important to be ready for how things in our industry are changing. In the hospitality world, there are lots of training programmes for areas like the Food & Beverage, the folks at the front desk, and the finance department. But there's not much out there for the technical side of things. That's why it's so important to make your own training programme for your team and/or yourself. The future is going to look different – we might start seeing lowering carbon emissions as part of annual bonuses, dealing with data is going to be part of the everyday job, and using advanced tools will be key. You want to be right there in the middle of it all, confident and ready for the changes that are coming. So, let's start thinking differently. Let's get ready for what you and your workplace need to be top-notch, making the most of your investments and help to maximise the turnover. That's the reason I put together my own training course for the team and why I'm suggesting you do the same. It's not the easiest path, but it's worth it. I use a small portion of my spare time to kick off a project called the Hospitality Engineering Development Programme. It got some good recognition quickly, setting up standards that I can take with me to any workplace. It's amazing how much can be achieved with consistency. Consistent actions, no matter how small, lead to significant results over time.

Now you know how to become a member of our community; "The key holders"!